Fed up with the right to food?

Fed up with the right to food?

The Netherlands' policies and practices regarding the human right to adequate food

edited by:

Otto Hospes

Bernd van der Meulen

Also available in the 'European Institute for Food Law series':

European Food Law Handbook
Bernd van der Meulen and Menno van der Velde
ISBN 978-90-8686-082
www.WageningenAcademic.com/foodlaw

ISBN 978-90-8686-107-1

First published, 2009

Wageningen Academic Publishers
The Netherlands, 2009

THE
RIGHT
TO
FOOD

F A O
FIAT PANIS

Table of contents

Abbreviations

AB	Administratiefrechtelijke Beslissingen (Review on administrative law case law)
ABRvS	Afdeling bestuursrechtspraak van de Raad van State (Administrative Judicial Review Division of the Council of State)
ACP-countries	Group of African, Caribbean and Pacific Countries
AKW	Algemene Kinderbijslagwet (social benefit law concerning family or child allowance)
Anw	Algemene Nabestaandenwet (social benefit law concerning income support in connection with the death of a family member)
AOW	Algemene Ouderdomswet (social benefit law: General Old Age pensions Act)
Awb	Algemene wet bestuursrecht (General Administrative Law Act)
CARICOM	Caribbean Community
CEDR	European Council for Agricultural Law (Comité Européen de Droit Rural)
CESCR	Committee on Economic, Social and Cultural Rights
CFS	Committee on World Food Security
COA	Centraal Orgaan opvang Asielzoekers (Central Organ for Reception Service for Asylum seekers)
CRvB	Centrale Raad van Beroep (Central Court of Appeal/Central Appeals Court; Dutch court specialised in civil servants law and social security law)
CVO	Civilian Volunteer Organization
EC	European Community
ECA	Export Credit Agency
ECHR	European Convention for the Protection of Human Rights and Fundamental Freedoms
ECnHR	European Commission on Human Rights
ECJ	European Court of Justice
ECR	European Court Reports
ECtHR	European Court of Human Rights
EHRLR	European Human Rights Law Review
EPA	Economic Partnership Agreement
ESC	European Social Charter
EU	European Union
FAO	Food and Agriculture Organization
FIAN	FoodFirst Information and Action Network
FIVIMS	Food Insecurity and Vulnerability Mapping System
G77	The Group of 77, consisting of 77 developing countries that are signatories of the 'Joint Declaration of the Seventy-Seven Countries'
GC12	General Comment 12
GFL	General Food Law

GMO	Genetically Modified Organism
GRULAC	Group of Latin American Countries
HACCP	Hazard Analysis and Critical Control Points
HR	Hoge Raad (The Dutch Supreme Court)
ICESCR	International Covenant on Economic, Social and Cultural Rights
ICRC	(International) Convention on the Rights of the Child
IFPRI	International Food Policy Research Institute
IGWG	Intergovernmental Working Group (For The Elaboration Of A Set Of Voluntary Guidelines To Support The Progressive Realization Of The Right To Adequate Food In The Context Of National Food Security)
ILO	International Labour Organization
IMF	International Monetary Fund
JB	Jurisprudentie bestuursrecht (Review on administrative law case law)
JV	Jurisprudentie Vreemdelingenrecht (Review on immigration law case law)
JWS	Johannes Wier Foundation
LJN	Landelijk Jurisprudentienummer (indication of case law published on the Internet on www.rechtspraak.nl)
MDG	Millennium Development Goals
MRLs	Maximum residue levels
NAPS	National Action Plan to Combat the Sexual Abuse of Children
NGO	Non Governmental Organization
NJ	Nederlandse Jurisprudentie (Review on civil and criminal law case law)
NJB	Nederlands Juristenblad (Dutch law journal)
NJCM	Nederlands Juristen Comité voor de Mensenrechten (Netherlands' Lawyers Committee for Human Rights/Dutch section of the International Commission of Jurists)
NJCM-Bulletin	Journal of the Netherlands' Lawyers Committee for Human Rights
ODA	Official Development Assistance
OECD	Organisation for Economic Co-operation and Development
OEWG	Open-Ended Working Group
OHCHR	Office of the United Nations High Commissioner for Human Rights
OJ	Official Journal of the European Union
PvdA	Partij van de Arbeid (Dutch political party)
RAwb	Rechtspraak Algemene wet bestuursrecht (loose leaf-collection of case law on the General Administrative Law Act)
RIA	Regulatory Impact Assessment
RSV	Rechtspraak Sociale Verzekeringen (Review on case law in social security)
RtF	Right to (Adequate) Food
SAP	Structural Adjustment Policy

SIM	Studie- en Informatiecentrum Mensenrechten. (Study- and information centre for human rights)
tB/S	ten Berge/Stroink (loose leaf-collection of case law in administrative law edited by Gio ten Berge and Frits Stroink)
TK	Tweede Kamer (second chamber of parliament, House of Commons)
TNC	Transnational Company
UDHR	Universal Declaration of Human Rights
UN	United Nations
UNESCO	United Nations Educational, Scientific and Cultural Organization
UNHCHR	United Nations High Commissioner for Human Rights
UNHCR	United Nations High Commissioner for Refugees
US	United States
USA	United States of America
USZ	Uitspraken Sociale Zekerheid (Review on social security law case law)
VG	Voluntary Guideline
WFS	World Food Summit
WFS:*fyl*	World Food Summit, five years later
WHO	World Health Organization
WSSD	World Summit on Sustainable Development
WTO	World Trade Organization
WWB	Wet Werk en Bijstand (Social benefit law concerning employment and income support)

About the authors

A.P.M. (Fons) **Coomans** occupies the UNESCO Chair in Human Rights and Peace at the University of Maastricht and is on the staff of the NJCM Bulletin. He was involved in the drawing up of the NJCM shadow report and attended the review of the Dutch government report in November 2006 at Geneva as an observer.

U. (Ute) **Hausmann** is staff member of the German section of the Food Information and Action Network and in this capacity representative in the Forum for Human Rights, the German network of human rights organizations.

O. (Otto) **Hospes** is Associate Professor of Law and Governance at Wageningen University. He is engaged in teaching and researching food (in)security, human rights and biofuels from a socio-legal perspective.

A. (Arne) **Oshaug** is Research director and Professor of public nutrition, Akershus University College, Norway.

B.M.J. (Bernd) **van der Meulen** is Professor of Law and Governance at Wageningen University (www.law.wur.nl). He is also chairman of the Dutch Food Law Association (www.nvlr.nl) and director of the European Institute for Food Law (www.food-law.nl).

F.M.C. (Frank) **Vlemminx** is an artist (painter). He is also Assistant Professor of Constitutional Law at Tilburg University (www.uvt.nl).

G.J. (Gijsbert) **Vonk** is Professor of social security law, Groningen University.

B.F.W. (Bart) **Wernaart** teaches law and ethics at Fontys University of Applied Sciences in Eindhoven and 's Hertogenbosch. He is also a professional musician (drums, mallets and percussion), composer and conductor.

Chapter 1

Introduction: what is there to celebrate in the Netherlands on World Food Day?

Bernd van der Meulen and Otto Hospes

1. World Food Day

The inspiration for writing this book on all relevant aspects of the Netherlands' policies and practices regarding the internationally recognised human right to adequate food was the celebration of World Food Day in 2007, or rather something that was ignored and not celebrated that day in the Netherlands.

Every year on 16 October, the anniversary of the Food and Agriculture Organisation of the UN, a 'World Food Day' is held to consider the world's food problems, both big and small, and to discuss improvements. Each year World Food Day is given a theme, around which, in principle, all 150 participating countries organise their activities. For 2007 the FAO chose 'the right to food' as the central theme.[1] In the FAO Council the United States of America argued strongly against this theme. The USA saw no reason to discuss hunger in the world within the framework of human rights issues. The Netherlands refrained from commenting in the FAO Council as regards the choice of theme, but neither did it give the theme a special place in the Dutch activities. The Dutch theme for 2007 was entitled 'The right to food or fuel'. Detailed and inspired discussions have taken place regarding the potential positive and negative effects of turning over agricultural land and wild life reserves to the production of fuel. The human rights perspective was only dealt with by a very limited number of speakers. When deciding upon the theme and particularly when deciding, despite the chosen title, not to put the human rights perspective in the limelight, undoubtedly an important consideration was the opinion expressed by the government's adviser, a professor in economics who e-mailed with regard to the human rights perspective: 'Otherwise I consider any opinion based on 'right' to be fundamentally wrong, whether this be the right to food or to fuel. Both, if taken seriously, boil down to a right to a minimum income. This right is recognised in the Netherlands but we contribute towards the cost of this together through politics (financial basis) and we have a social fiscal number (administration). As a means to combat poverty and hunger worldwide it is for the present nonsensical and misleading. Apart from that in the past the right to food has been established a few times in connection with the UN without anyone taking it seriously.'

[1] www.fao.org/wfd2007/index_en.html.

However interesting and well organised the Dutch programme was, surely it can not be true that the whole world was discussing the right to adequate food, except for the Netherlands? The more so when it appears there is also a lack of understanding and interest amongst experts. Even if one were to conclude that it is 'nonsensical and misleading', we should not keep quiet about it.

2. World Hunger Day

The official programme of the Netherlands on World Food Day triggered FIAN Netherlands[2] and the chair group of Law and Governance at Wageningen University – in cooperation with the European Institute for Food Law – to organise a one-day conference during which the human rights approach was given centre stage. This alternative Dutch celebration of World Food Day was labelled 'World Hunger Day'. The initiative attracted wide support. Speakers came from FIAN-International, the NJCM (the Netherlands Committee of Jurists for Human Rights), the Labour Party (Partij van de Arbeid), the Netherlands Food Bank Foundation, the Free University of Amsterdam and the Universities of Groningen, Tilburg and Akershus (Norway).[3] The meeting took place on 15 October 2007, the day before World Food Day, in the building of the Institute for Constitutional and Administrative Law at Utrecht University. The question of how, from both a national and international perspective, the Netherlands approaches the right to adequate food in its law and its policy was discussed from different points of view.

The choice of subjects dealt with in this book is based on this conference. Some chapters are elaborations by the speakers of the presentations they held, some have been specially written for this book by others. Each chapter deals with the Netherlands' policies and practices at the national or international level. These chapters are preceded by a chapter written by Van der Meulen and Vlemminx based on the report on the Netherlands they wrote for the 2005 CEDR conference on the Right to Food. This chapter sets the stage on which the other contributions present all relevant developments after 2005. We will now introduce the right to adequate food and the chapters of this book on the basis of a pre-view of how the Netherlands deals with this right domestically and in international affairs.

[2] FIAN stands for FoodFirst Information and Action Network. FIAN is an international human rights organisation that campaigns for the right to food. See www.fian.org.

[3] For the programme and background documentation see: www.food-law.nl.

3. The right to adequate food

The right to food, FIAN prefers to refer to the right to (be able to) feed oneself, is recognised in different international humanitarian and human rights conventions.[4] The core provision is Article 11 of the International Covenant on Economic, Social and Cultural Rights (ICESCR).

Professor Oshaug remarked that it was a bit late to be asking whether the right to adequate food is binding law in the Netherlands. The Netherlands has signed and ratified the ICESCR. It entered into effect in the Netherlands on 11 March 1979. You cannot sign first and then say you will have no more to do with it. Article 11 ICESCR reads as follows:

1. The States Parties to the present Covenant recognize *the right of everyone* to an adequate standard of living for himself and his family, including *adequate food*, clothing and housing, and to the continuous improvement of living conditions. The States Parties will take appropriate steps to ensure the realization of this right, recognizing to this effect the essential importance of international co-operation based on free consent.
2. The States Parties to the present Covenant, recognizing the fundamental right of everyone to be *free from hunger*, shall take, individually and through international co-operation, the measures, including specific programmes, which are needed:
   a. To improve methods of production, conservation and distribution of food by making full use of technical and scientific knowledge, by disseminating knowledge of the principles of nutrition and by developing or reforming agrarian systems in such a way as to achieve the most efficient development and utilization of natural resources;
   b. Taking into account the problems of both food-importing and food-exporting countries, to ensure an equitable distribution of world food supplies in relation to need.'

How exactly should this text be interpreted? Over the years people have worked hard to concretise the right to adequate food. Further to questions referred to it and on the basis of experience in the various UN member states, the UN Committee on Economic, Social and Cultural Rights has written 'General Comments'(GC) on the different provisions in the ICESCR. GC 3 concerns the obligations of the member states and GC12 concerns the right to adequate food. UN Special

[4] Annex 1 gives the most important treaty provisions on the human right to food. Annex 2 includes two relevant General Comments. Annex 3 contains the FAO Voluntary Guidelines on the Right to Food. Together these annexes form the international body of law on the right to food. In many UN declarations and conventions the treaty provisions have been confirmed. Outstanding examples are the UN Millennium Declaration, the UN Women's Convention, the UN Convention on the Rights of the Child and the Geneva Conventions. The article of Hospes (2008) on 'Overcoming Barriers to the Implementation of the Right to Food' as published in the European Food and Feed Law Review, provides an overview and analysis of the international body of law on the right to food.

Rapporteurs report on the situation regarding the right to food.[5] After and further to the discussions at the World Food Summit five years later (2002) an FAO working group consisting of representatives from all continents produced guidelines for the concrete application of the right to adequate food in the FAO member states: the so-called *'Voluntary Guidelines to support the progressive realisation of the right to adequate food in the context of national food security'*.[6] The right to adequate food is realised if people have access to food that:

- provides sufficient nutritional value and micronutrients for a person to lead a healthy and active life;
- is free of hazardous substances;
- is acceptable within a given culture.

The human rights approach can, in other words, be adopted as a point of departure for both food security and food safety.

Rights always go hand in hand with obligations. Human rights go hand in hand with state obligations. In the GCs and the Voluntary Guidelines three types of obligations are distinguished:

1. The obligation *to respect*. In general people are able to care for themselves and their families. This ability may not be curbed without sound legal justification; this is in line with other fundamental rights such as the freedom of expression for instance.
2. The obligation *to protect*. If the ability of citizens to provide for themselves is threatened by other citizens the government must do its best to protect these citizens from the others.
3. The obligation *to fulfil*. This obligation is composed of a policy obligation and a relief obligation. On the one hand a prudent government is expected to adopt policy geared towards supporting and promoting the ability of the population to provide for itself, on the other hand it must do its best to provide assistance if people find themselves in a situation in which they cannot provide for themselves through no fault of their own.

Sometimes these obligations require further elaboration in legislation and (economic) policy. Sometimes they are deemed to be hard and solid enough for the courts to apply them directly in concrete cases. Examples of such direct applicability can be found close to home (Belgium, Switzerland) as well as further away (South Africa, India).[7] States that are party to the ICESCR are expected to report on the progress made by them periodically (every 5 years). The next

---

[5] For the most relevant UN documents see: www.unhchr.ch/huridocda/huridoca.nsf/FramePage/Subject+food+En?OpenDocument.

[6] See: www.fao.org/docrep/meeting/009/y9825e/y9825e00.htm.

[7] Regarding these countries see: Wenche Bart Eide en Uwe Kracht (red.) Food and Human Rights in Development. Vol. II Evolving Issues and Emerging Applications, Intersentia Antwerpen 2007.

section provides some glimpses of the situation in the Netherlands in policy and in practice as regards respecting the human right to adequate food.

4. The Netherlands domestically

4.1 Poverty in the Netherlands

In theory the Netherlands has a respectable social security system. In practice, not everyone can keep his or her head above water. For many the red tape is too complex. Those with no permanent address or place of residence, find it hard to gain access to municipal facilities. People with mental problems don't have an easy time of it. The subject of poverty has caught the attention of both the Dutch Central Bureau of Statistics and the Socio-Economic Planning Bureau. Every other year they publish a joint detailed report by the name of 'Armoedemonitor' (Poverty Monitor) which reflects the state of affairs in cold statistics. In the even years they publish a slightly less detailed 'Armoedebericht' (Poverty Bulletin). The Poverty Monitors of 2005 and 2007 and the Poverty Bulletin of 2006 show that some 10% of households in the Netherlands (some two thirds of a million) live below the poverty line.[8]

4.2 Food Banks

The plight of the hungry in the Netherlands has attracted private initiatives. In 2003 the first food bank was founded in Rotterdam. There are currently some 140 food banks in the Netherlands. Jan Willem Reineke of the Dutch Food Bank Foundation shared his experiences at World Hunger Day. The food banks put together food parcels consisting of food products donated by the business community. Anyone who asks is entitled to receive one parcel once. Anyone wishing to receive a second parcel however first has to attend an intake interview. During this interview the person's situation is examined and ways in which he or she can be helped out of this situation are looked at. The food parcels are a means not an end in themselves.

Initially it was suggested that World Hunger Day should take place in a food bank. The choice of location alone would immediately make clear the relevance of the theme. However the suggestion was considered problematic, in particular by customers of the food banks. Dependency on food banks is embarrassing for many and for this reason they prefer as few outsiders as possible to witness this dependence.[9]

[8] See for more details www.armoedemonitor.nl, the official website.
[9] For further information on the food banks in the Netherlands see www.voedselbank.nl. See also in the Dutch language the Master's thesis by Hille Hoogland 'Voedselhulp in een land van overvloed' 2006, available on the webpage of World Hunger Day.

4.3 Hunger as a policy instrument?

The policy adopted in the Netherlands in respect of failed asylum seekers and (other) illegal immigrants in the early 1990s (in other words many years before Rita Verdonk had her sobering influence on Dutch immigration policy!) was initially also referred to as 'smoking out' but is now called 'linking'. Under this policy rejected immigrants were no longer able to make use of the 'bed and bread' scheme in force at that time nor were they granted access to (any other) social facility. The only thing they could expect from the Dutch government was a return ticket to their country of origin. The current belief, not expressed quite as emphatically as in the beginning, is that the withdrawal of all facilities encourages people to leave the country. In this sense it can be seen as a policy instrument.

Gijs Vonk, professor at the VU (Free University, Amsterdam) and the RUG (University of Groningen), sees things in a slightly different and wider perspective. In social security law there is an increasing tendency to use the withdrawal of rights as a sanction for failure to fulfil obligations. According to Vonk this does not as a rule involve people who consciously commit fraud but people who lose their way in the maze of red tape.[10]

The threat associated with the risk of having assistance withdrawn is undoubtedly a powerful stimulant to fulfilling one's administrative obligations to the best of one's ability. The question of whether the actual withdrawing of assistance from people who make mistakes is a justifiable limitation of human rights such as the right to adequate food is however not asked in the Netherlands, let alone answered. Chapter 5 of this book provides the full argument of Vonk on the issue of hunger as a policy argument.

4.4 Starving for (case) law

The further details provided by the UN with regard to the obligations of states that are party to the ICESCR and the calls made upon the domestic courts to ensure compliance with these obligations have, to date, been systematically ignored by the Dutch courts. The Netherlands continues to uphold the most outdated opinion that differentiates between classic fundamental rights (civil and political rights) that impose negative (or passive) obligations (or obligations not to take measures) upon the government that can be enforced by the courts and social fundamental rights that impose positive (active, or performance) obligations upon the government that should be interpreted as being policy objectives and not as enforceable legal obligations.

[10] He further elaborated upon this theme in his inaugural speech held on 29 January 2008 at Groningen University.

A shocking example is presented by the Amsterdam Court of Law (13 March 2001) in a case regarding the application of the 'smoking out' policy.[11] A failed asylum seeker from Somalia from whom all facilities were withdrawn, appealed, among other provisions, to Article 11 ICESCR. According to the Court however this was not a self-executing provision because the term 'adequate food' is too vague. After all where should one draw the line between adequate and no longer adequate? Frank Vlemminx (University of Tilburg) has been highly critical of this and similar rulings in a number of publications. The question of where to draw the line is totally irrelevant in a situation where there is nothing at all. 'Nothing' is always too little. However this time he did not emphasise the embarrassing failures in Dutch case law but the opportunities being created by the European Court of Human Rights to give weight to socio-economic human rights via the civil and political rights. The Court only has jurisdiction to apply provisions in the European Convention for the Protection of Human Rights and the Fundamental Freedoms (ECHR) and not provisions in other human rights conventions such as the European Social Charter (ESC) and the ICESCR. However, when elaborating upon the rights pursuant to the ECHR the Court is increasingly taking the other conventions into account. When a fundamental right is involved for example, this gives an extra dimension to the application of the discrimination prohibition. Vlemminx refers to this as convergence. Vlemminx advises lawyers to place complaints involving a violation of socio-economic rights as much as possible within the context of classic fundamental rights.

In Chapter 3 of this book Vlemminx argues that the Netherlands is hungry for (case) law. He explains why Dutch constitutional law is schizophrenic and why the case law of the European Court for Human Rights could be an example for the Dutch. In Chapter 4 Wernaart shows, however, that some precedents in Dutch case law can be found that are relevant for people in need.

5. The Netherlands internationally

5.1 Reports

Pursuant to Article 16 and 17 ICESCR the member states have to report on the progress they have made in realising the rights embodied in this Covenant. The Netherlands submitted its *third* report on 18 August 2005. More than eight years late. In fact its *fourth* report should have been submitted in 2002. The reports submitted by the Netherlands to the UN Committee on Economic, Social and Cultural Rights are each time shamelessly late and brazenly rose-tinted.

Maria Lourijsen (NJCM) was involved in the shadow reporting. The UN Committee on Economic, Social and Cultural Rights gives NGOs the opportunity to respond

[11] Published in RAwb 2001, 74 m.n. Vlemminx. Also on the web page of World Hunger Day.

orally or in shadow reports to the official reports and to draw up questions. NJCM and the Johannes Wier Foundation made use of both opportunities. Their general comment on the Netherlands' third report is that the report is highly descriptive and rose-coloured. The government is supposed to report on the implementation of economic, social and cultural rights in the Netherlands but barely touches upon potential obstacles such as those mentioned in Article 17 of the Covenant. The Netherlands puts these rights 'out of play' by stating in its third report that 'most of the provisions in the ICESCR are not directly applicable'.

In Chapter 7 of this book the issue of 'rosy reports' from the Netherlands is critically reviewed by Coomans.

5.2 De FAO Voluntary Guidelines

In 1996 the Millennium Development Goals were formulated during the World Food Summit. One of the most important goals is to halve the world's hunger before 2015. Five years later during the World Food Summit in 2002 an interim balance was made. The progress was nothing to write home about. One of the initiatives that was undertaken to make the states see the urgency of the action was the setting up of an intergovernmental FAO workgroup to come up with recommendations relating to the way in which the right to food can be realised. Arne Oshaug represented Norway in the negotiations. During its EU presidency the Netherlands represented the EU in the intergovernmental working group.

The negotiations met two major stumbling blocks. The first was whether binding law or voluntary guidelines should be applied. The second major stumbling block was the international dimension. This concerned the question of whether or not states' obligations in respect of other countries and their citizens should be referred to in the recommendation. Neither the Netherlands nor the US was in favour of binding law or reference to the international obligations. Their attitude was decisive. In the end the working group formulated voluntary guidelines and assembled existing international obligations in a separate section of the recommendation but did not add to these.

Chapter 6 of this book provides an interesting account and insider's view of Oshaug of the Netherlands' position and role in the making of the voluntary guidelines on the right to food.

5.3 Extra territorial obligations

Article 11 ICESCR explicitly calls upon the member states to cooperate at an international level. The question of exactly how voluntary or obligatory this call should be considered is the subject of discussion in many countries. Does the right to adequate food have extra territorial effect? Ute Hausmann of FIAN International discussed a number of practical situations that demonstrate how choices made in one country can cause famine in another country.

In connection with this FIAN international asked the Inter American Court for Human Rights to test a bilateral investment convention concluded between Germany and Paraguay. An important clause in the convention was that German landowners could not be dispossessed. The court decided that this contravened the human right to food and ruled that Paraguay is permitted to dispossess German land owners and turn the land over to agricultural food production.

FIAN International also focuses its actions on international funding institutions, such as the International Finance Corporation (IFC). This funding agency wanted to make a loan of USD 125 million available to mine works that would contravene rights to land and food. FIAN was able to convince the German member of the IFC board of the risk that human rights would be contravened. The loan still went through because the other members of the board raised no objections.

To make extra territorial obligations more effective, a number of obstacles need to be addressed, such as the weak legal basis upon which victims in poor countries can claim their rights *vis-à-vis* bilateral or multilateral aid workers, credit providers and investors abroad; the failure of rich countries to recognise victims in poor countries as rightful claimants; and the absence of objection procedures at international institutions. According to Hausmann regional human rights bodies could form an important opening for making extra territorial obligations more effective by conducting concrete case studies. However, their geographic limitations make it necessary to define and elaborate upon extra territorial obligations in a UN connection.

In Chapter 9 of this book Ute Hausmann explains that extra-territorial obligations provide opportunities and legal grounds for the hungry to challenge the global elite. In particular she discusses FIAN's struggle regarding the acceptance of international obligations by Germany. To date this kind and level of discussion has not taken place in the Netherlands. As shown by Oshaug in Chapter 6, the Netherlands is one of the countries trying to prevent the acceptance of extra territorial obligations.

5.4 The right to food or fuel? The tense relationship between food, biofuel and human rights

In his argument on biofuels and the right to fuel Otto Hospes suggested that the increasing production and use of crops for energy production forms a major threat to world food security. Worldwide the production of ethanol and biodiesel tripled between 2000 and 2005 and this seems to be only the beginning. Since 2005 many national governments have formulated a biofuel policy aimed at the mandatory or voluntary mixing of biofuels with fossil fuels for transport and electricity. Multinational enterprises in agriculture, trade, energy and banking are investing heavily in the biobased economy.

With a view to minimising the potential negative effects of biofuel expansion a project group, headed by the then professor Jacqueline Cramer (now: Minister for the Environment) presented the Dutch government with recommendations in April 2007. These recommendations contain principles, standards and criteria for the socially, ecologically and economically sustainable production of biomass, regardless of whether this biomass is produced within Europe or without. It is remarkable, according to Otto Hospes, that principles have been suggested with regard to food security and combating poverty but no criteria, never mind standards, have been proposed. Fortunately this is not the case with regard to the social welfare theme. Two of the criteria stated in connection with this are that biomass production may not result in human rights or property rights being violated.

By formulating the recommendations in this way the Cramer project group, composed of ministers, members of the business community and civil society, not only produced a coherent vision on sustainability but in doing so they also stated human rights (including the right to food!) as being an integral part. The project group's recommendations hereby present the government with the chance to make the ICESCR, signed by the Netherlands, more effective in new policy fields. However, Hospes questions whether the Dutch government will be able to handle this and defy the anticipated resistance during talks on this issue at EU and WTO level. A difficult point regards the extent to which the setting of criteria in respect of the import of biofuels in terms of human rights is compatible with WTO legislation that in principle allows no restrictions to free trade. Does the Netherlands dare risk bringing a case down on its head? Or will the Netherlands decide to leave the matter to the business community which, together with sections of civil society, will set the (private) standard?

In Chapter 8 Hospes reviews the pioneering role of the Netherlands in formulating sustainability criteria for biofuel production. He shows that the Netherlands is in want of sustainable biofuels but considers it unnecessary to take the right to food and human rights in general as the point of departure or normative framework for Dutch international policies and cooperation in the field of biofuels.

6. Fed up with the right to food?

A central objective of World Hunger Day was to discuss whether the Netherlands was correct in assuming that the right to food is not a relevant concern for the Netherlands and therefore has neither a place within the Netherlands celebration of World Food Day nor on domestic and international policy agendas of the

Netherlands. We think that this assumption is wrong, both for the domestic and international scene.

There is probably no Dutch man or woman who has not seen a fellow countryman rummaging through a dustbin for food. Poverty and hunger have become a reality; fortunately not for many, but this makes it no less distressing for those who are affected by it. Food banks help out where the government fails to honour its responsibilities. The government uses the dependence of some people on social facilities as a means of coercing these people to comply with their obligations. The question of whether this contravenes human rights is not addressed. The courts keep the socio-economic human rights, including the right to food, systematically at bay. A great deal of judicial dexterity is required to be able to achieve anything with these rights.

Calculated per head of the population, the Netherlands is one of the richest countries in the world. Pro rata it should therefore be relatively easy for the Netherlands to comply fully with the right to food. If in this privileged situation the Netherlands was more willing to apply the right to food, the Netherlands could make a valuable contribution to the development of this right. Experience in the Dutch situation could benefit people and governments in more difficult times or parts of the world by setting the example and by answering questions regarding the reach of state obligations, situations where claims can be made, justifications for limitations to the right to food and all other issues that are commonly resolved when human rights are recognised as living law.

The Netherlands is grossly neglecting its reporting obligations. With shameless delay it states that all is well. In the international arena the Netherlands is one of the countries pulling on the hand brake. During the negotiations on the FAO's guidelines the EU, under the Dutch presidency, (together with the US) made every effort to ensure that the guidelines were no more than 'voluntary' in character and contain no enforceable obligations relating to human rights. External aspects were also radically reduced. The practice shows us that it is these very external situations in which countries can make a difference. Is the Netherlands ready to make a difference? The policy in respect of biofuels is one area in which this question can be asked, but not the only area.

Both in national and international debates the Dutch government, represented by a number of cabinets has been hammering home the importance of norms and values. Unfortunately our government does not appear to be very clear about what these norms and values actually comprise. The first Balkenende cabinet even asked for advice on this matter. Is it not obvious that first and foremost we should look for our norms and values in our own Constitution and in international conventions on human rights to which the Netherlands is party?

Chapter 2
An adequate right to food?
The Netherlands: abundant in food, wanting in law

Prof. dr. B.M.J. van der Meulen and dr. F.M.C. Vlemminx[12]

1. Introduction

This chapter gives an account of the right to food in Dutch law initially written for Commission I of the 2005 CEDR conference in Røros, Norway.[13] It has been redrafted to provide a general background to the other chapters in this book each focussing on a more specific topic.

After this introduction, this chapter continues in Section 2 with some background information on the legal system of the Netherlands in general and the role of human rights in particular, including some information on human rights education in the Netherlands. Section 3 is dedicated to the place of Article 11 of the International Covenant on Economic, Social and Cultural Rights (ICESCR), the right to adequate food, in Dutch law. It will be seen that the impact of Article 11 ICESCR is limited. For this reason some remaining subjects have been treated outside the scope of Article 11 ICESCR in Section 4. This part of Dutch food law consists mostly of directly applicable EU law and national legislation harmonised by EU directives. In Section 5 a short conclusion is drawn.

The text has not been substantially updated after 2005 because other chapters in this book provide the relevant information on subsequent developments.

2. Background

2.1 The Dutch legal system

In this chapter we will refer to Dutch case law and parliamentary proceedings. This approach follows from the Dutch system of law. For a better understanding of our reasoning we will start with some background information concerning this system of law.

---

[12] The order is strictly alphabetic. The authors contributed in equal measure to this chapter. Many thanks for the input of Aaron Chase Underwood (at the time of writing, student of International Law from Topsham, Maine, USA) and Jaap S. Kloet (at the time of writing, student in Master's Intentional Development Studies at Wageningen University). Comments are welcome at: Bernd.vanderMeulen@wur.nl.

[13] For the reports on other countries, see www.cedr.org.

In comparative law the Netherlands[14] is included with the civil law family of continental Europe.[15] Among other things, this means that legislation is the most important source of law.[16] The Netherlands has a written constitution[17] to which lower legislation must conform. The courts can annul legislation that departs from the constitution. There is one exception to this rule. The highest level of legislation, called 'formal legislation' or sometimes translated as 'Acts of Parliament', is legislation forged in co-operation between the national government and parliament, and may not be annulled for unconstitutionality. The idea is that the formal legislator with its democratic basis has the power of final interpretation of the Constitution, not the courts. In all other cases the courts have the power of 'authentic interpretation'. That is to say, the courts have the final say on what legislation means, if it is valid, and more generally what the law is. This gives case law a status almost equal to legislation as a source of law. Although case law exercises considerable influence, there is no doctrine of binding precedent. In theory, the courts are free to depart from earlier decisions given at the same or even higher levels. In practice, however, they will be reluctant to do so in cases of established case law. For this reason case law is a key factor in legal analysis.

There are four courts that are highest in their respective field(s) of law. The 'Hoge Raad' ('Supreme Court') is the 'Court de cassation' in civil, criminal and fiscal cases. It judges on matters of law, not of fact. The three other highest courts are all administrative courts that pass judgements both on the law and on the facts. They are the 'College van Beroep voor het bedrijfsleven' ('Industrial Appeals Board'[18]) specialising in economic administrative law. It will not be mentioned again in this chapter; the 'Centrale Raad van Beroep' ('Central Court of Appeal') specialising in civil servants law and social security law; and the 'Afdeling bestuursrechtspraak van de Raad van State' ('Administrative Judicial Review Division of the Council of State') for all other administrative law cases.[19]

---

[14] Adjective: Dutch. After its seafaring provinces on the western cost the Netherlands is also known as 'Holland'.

[15] For an introduction to Dutch law in English see: Ewoud Hondius, Jeroen Chorus (eds.) Introduction to Dutch Law, 4th ed. Kluwer Law International 2006; Sanne Taekema (ed.) Understanding Dutch Law, BJu (www.bju.nl) The Hague 2004; Leonard F.M. Besselink, Constitutional Law in the Netherlands, Ars Aequi Libri Nijmegen, 2004; see also Oswald Jansen and George Middeldorp, Researching Dutch Law www.llrx.com/features/dutch.htm#top). A sociological portrait of Dutch law is available on the Internet, Fred J. Bruinsma, Dutch Law in Action: www.uu.nl/content/dutchlawinaction2003.pdf.

[16] But it is not necessarily the highest. In Dutch doctrine four sources of law are usually distinguished: international treaties (including secondary international law), written legislation, case law, and unwritten law (like general principles of law, general principles of good governance and customary law).

[17] For an English translation see www.minbzk.nl/contents/pages/6156/grondwet_UK_6-02.pdf.

[18] Known from ECtHR 19 April 1994, case no. 16034/90 (Van der Hurk case; the Commission decided on 10 December 1992, no. 16034/90), and in EC law from many preliminary questions.

[19] Before 1994 it was called the 'Afdeling rechtspraak van de Raad van State'; *Judicial Division of the Council of State.*

The courts use a variety of instruments for the interpretation of legislation. Important among these is the so-called 'legal-historic' method. When they apply this method the courts rely on the intentions of the legislators as they appear in the published proceedings of parliament.

The Dutch Constitution has a monistic approach to international law, as explained below.

2.2 Human rights in the Dutch Constitution

The Dutch Constitution was re-codified in 1983. The new Constitution opens with a chapter on fundamental rights ('grondrechten'). This chapter contains civil and political rights as well as economic and social rights. In Dutch legal doctrine the old-fashioned distinction between justiciable fundamental freedoms that basically require non-intervention from the authorities and non-justiciable economic, social and cultural rights that require positive action from the authorities, is still alive. Most economic and social rights are set out not as entitlements of citizens but as concerns of the authorities.

More subtle distinctions between different kinds of negative and positive obligations connected to all human rights (e.g. obligations to respect, to ensure, to promote and to protect) are present in literature as well,[20] but are not prevalent.

2.3 The right to food in the Dutch Constitution

The right to food as such is not set out in the Dutch Constitution. The government found it unnecessary to mention it separately as it was considered to be implicit in the right to means of subsistence laid down in Article 20.

Article 20 of the Constitution of the Netherlands

1. It shall be the concern of the authorities to secure the means of subsistence of the population and to achieve the distribution of wealth.
2. Rules concerning entitlement to social security shall be laid down by Act of Parliament.
3. Dutch nationals resident in the Netherlands who are unable to provide for themselves shall have a right, to be regulated by Act of Parliament, to aid from the authorities.

This low-key approach is somewhat surprising. Discussion on human rights in the Netherlands was strongly influenced by the traumatic experience of the German

[20] Mainly in the writings of Frank Vlemminx, like Een nieuw profiel van de grondrechten. Een analyse van de prestatieplichten ingevolge klassieke en sociale grondrechten, third edition, BJu The Hague 2002 and De autonome rechtstreekse werking van het EVRM. De Belgische en Nederlandse rechtspraak over verzekeringsplichten ingevolge het EVRM (preadvies Nederlandse Vereniging voor Rechtsvergelijking) Kluwer Deventer 2002.

occupation during World War II. A substantial part of the population suffered during the infamous 'winter of famine' of 1944. A few decades later food as an issue of human rights discussion seems to have been all but forgotten.

2.4 Human rights education in the Netherlands

All law faculties in the Netherlands teach courses in Constitutional Law of which human rights law is an important part. The European Convention for Human Rights and Fundamental Freedoms is treated in a variety of courses ranging from administrative to civil and criminal law. The reason is that the Dutch courts often apply this convention. Also, in International Law, attention is given to the UN and to human rights. Usually, however, ICESCR is mentioned only in passing.

There is a graduate school devoted to human rights. Several universities work together in this graduate school, the secretariat of which is at Utrecht University.[21] SIM, the Documentation Centre Netherlands Institute of Human Rights[22]also resides at Utrecht University. Outside the Universities, the T.M.C. Asser Instituut[23] (in The Hague) is involved in the study and promotion of human rights.

NJCM[24] publishes a journal (NJCM-Bulletin) on all aspects of human rights. There is also the Netherlands Quarterly of Human Rights in English language.[25]

At present Wageningen University is the only university in the Netherlands that teaches Food Law. The right to food is treated in this context.[26]

3. The right to adequate food in Article 11 ICESCR

3.1 Introduction

The Netherlands is a state party to the ICESCR. The Covenant took effect in the Netherlands on 11, March 1979, a little over three years after it entered into force on 3, January 1976 and more than twelve years after it was concluded.

[21] Website: www.uu.nl/uupublish/homerechtsgeleer/onderzoek/onderzoekscholen/rechtenvandemens/english/20494main.html.
[22] Studie- en Informatiecentrum Mensenrechten. See: http://sim.law.uu.nl/SIM/Dochome.nsf?Open.
[23] Website: www.asser.nl.
[24] Nederlands Juristen Comité voor de Mensenrechten; Netherlands' Lawyers Committee for Human Rights (www.njcm.nl/site/).
[25] See: www.intersentia.be/magazine.aspx?magazineId=PUB10007.
[26] For the Wageningen approach to food law see: B.M.J. van der Meulen, The Right to Adequate Food. Food Law Between the Market and Human Rights (inaugural address) ISBN 90 5901 9628, Elsevier 2004, and B.M.J. van der Meulen and M. van der Velde, European Food Law Handbook, Wageningen Academic Publishers, 2008 (www.wageningenacademic.com/foodlaw).

In the Netherlands, the fate of all rights in the ICESCR ('the Covenant'), including the right to food contained in Article 11, has been intertwined from the outset with that of the equivalent rights in the Dutch Constitution. Under Article 91 of the Constitution a treaty must be approved by Act of Parliament. When the Covenant was submitted for approval to Parliament in the mid-seventies, the government in the explanatory memorandum on the bill described the nature of the Covenant as follows, using Article 2 of the Covenant as an example:[27]

> *'The first paragraph clearly shows that the drafters of the Covenant have considered that the economic, social and cultural rights in terms of nature and contents because of their generality do not offer a fixed standard for the rate and degree of realisation of those rights.'*

In the government's opinion the Covenant solely contained open standards whose contents should be defined by politics and policy. In particular with regard to Article 11 of the Covenant the government pointed out:[28]

> *'that in particular in developing countries the living conditions of large population groups are far below an acceptable level and that in those countries the realisation of ever-improving living conditions is an urgent necessity for many years to come.'*

During virtually the same period Parliament was considering the inclusion of economic, social and cultural rights in the Dutch Constitution and the government provided those rights with similar comments. From the wording of the constitutional articles in question:[29]

> *'an ample policy margin has been left for the bodies entrusted with regulations or care. The policy margin means primarily that the rate and speed of realisation of the interests phrased in the provisions are left to the government body concerned.'*

The government further took the position that the implementation of the economic, social and cultural rights mentioned in the Constitution had been almost realised.[30] For this reason the government dismissed the judicial control of compliance. Only in very exceptional cases would the Court be able to establish a violation of the rights. The government considered the possibility that such a case would arise as

[27] Parliamentary Documents II 1975-1976, 13 932 (R 1037), no. 3, page 45.
[28] Parliamentary Documents II 1975-1976, 13 932 (R 1037), no. 8, page 26.
[29] Algehele grondwetsherziening (*Overall constitutional revision*), Deel Ia Grondrechten, Den Haag 1979, page 258.
[30] Algehele grondwetsherziening (*Overall constitutional revision*), Deel Ia Grondrechten, Den Haag 1979, page 255.

virtually academic.[31] In brief, the government lumped together the rights contained in the Covenant and the Constitution. That the wording of the rights contained in the Constitution was extremely vague where the Covenant was rather detailed on many points escaped notice or was considered irrelevant.

3.2 The rights contained in the Constitution and in the Covenant in legal practice

The government's position on the justiciability of the economic, social and cultural rights contained in the Constitution has had a huge effect on legal practice. Since the rights were embedded in the Constitution in 1983 the courts have established just two violations.[32] In addition the National Ombudsman has held once that a right was being violated.[33] This does not mean that the government carefully honours the rights, for in almost all cases the courts dismiss evaluation against the rights with a reference to the government. As will be shown below the rights contained in the Covenant do not fare any better.

Of course the significance of the economic, social and cultural rights contained in the Constitution does not depend exclusively on the possibility of judicial control. It is conceivable that the rights are incorporated into legislation and administration entirely voluntarily. However, an extensive evaluation study has been performed, which showed that the economic, social and cultural rights contained in the Constitution play absolutely no role in the legislation process and administrative actions.[34] More recent, but less extensive studies point in the same direction.[35] Although the significance of the Covenant for legislation and administration was not studied separately, it is safe to assume that this gloomy picture applies just as much to the rights contained in the Covenant.

3.2.1 Direct applicability of the Covenant

The answer to the issue of judicial control of compliance with the Covenant should first of all consider the manner of reception of the Covenant's provisions. The Netherlands applies a monistic system. This was decided by the Supreme Court as early as 1919.[36] This means that within the Dutch judicial system treaty

[31] Algehele grondwetsherziening (*Overall constitutional revision*), Deel Ia Grondrechten, Den Haag 1979, page 258.
[32] Summary proceedings President of the Judicial Division of the Council of State, 10 May 1989, AB 1989, 481 on the right to education in Article 23 Constitution and Court of Utrecht, 18 June 1991, NJ 1992, 370 on the right to housing in Article 22 (2) of the Constitution.
[33] No 18 July 1995, report no. 95/271 on the right to health of Article 22 (1) Constitution.
[34] Tj. Gerbranda and M. Kroes, Grondrechten evaluatieonderzoek, final report, Leiden 1993, page 334.
[35] F. Vlemminx, Een nieuw profiel van de grondrechten, The Hague 2002, pages 20-22.
[36] Supreme Court, 3 March 1919, NJ 1919, page 371; F.M.C. Vlemminx and M.G. Boekhorst, Article 93, in: A.K. Koekkoek (ed.), De Grondwet, Een systematisch en artikelsgewijs commentaar, Deventer 2000, pages 456-457.

provisions can be invoked if they have direct applicability. The Netherlands, however, applies a special term in connection with this direct applicability. This is the concept 'binding on all persons' that was included in Article 94 of the Constitution in 1956. It is a rather obscure phrase, and besides the question as to whether the provisions of the Covenant award discretionary powers to the legislator and/or administration, other factors play a role as well.[37] These other factors aim at evading direct applicability.

First of all, especially when the Covenant is at stake, the courts place great value on the position taken by the government in its explanatory memorandum on the approval act about direct applicability. In view of earlier remarks it will come as no surprise that the government dismisses direct applicability of the Covenant:[38]

> *'In general the provisions contained in this Covenant will not have direct applicability.'*

When the Covenant is invoked the courts usually refer to this comment.

A second characteristic of the phrase 'binding on all persons' implies that direct applicability may require that the provision of the Covenant addresses citizens directly. The word 'may' already shows that this requirement is not always set. However, partly for this reason direct applicability of the Covenant is often dismissed.

Especially with regard to the Covenant a third characteristic of the phrase 'binding on all persons' is that direct applicability is assessed in an extremely abstract manner. The case to be resolved by the Court does not play any role in the considerations. For instance, direct applicability of the right to adequate food contained in Article 11 of the Covenant may be denied because in the Court's view the term 'adequate food' is too vague, while in the case at issue *no food at all* is supplied and Article 11 is undeniably violated.[39]

This specific approach to direct applicability of the Covenant implies that the General Comments of the Committee on Economic, Social and Cultural Rights on the assessment of direct applicability appear to play no role at all. No matter how much the Committee emphasises that the policy freedom is restricted or even absent, this does not make a difference for the Dutch courts with regard to direct applicability. It is not the issue of policy freedom but the other factors mentioned above that are decisive. Even the discussion about direct applicability

---

[37] See about this subject: F. Vlemminx, Een nieuw profiel van de grondrechten, The Hague 2002, page 207-210.
[38] Parliamentary Documents II 1975-1976, 13 932 (R 1037), no. 3, page 13.
[39] Court of The Hague, 6 September 2000, RAwb 2001, 55.

in General Comment 3 on the nature of States parties obligations and General Comment 9 on the domestic application of the Covenant do not change this. The effect seems negligible, even zero. The same applies to the typology of obligations, whose relevance has been stressed for years by Dutch legal scholars.[40]

3.3 Case law of the Central Appeals Court

During the 25 years that the Covenant has been in force in the Netherlands, the courts have established a violation just once. This concerned a cutback measure that affected the right to equal remuneration (Article 7 of the Covenant) of public officials.[41] This ruling was set aside by the Central Appeals Court five years later, in 1989, for reason of the absence of direct applicability.[42]

Not only is the Central Appeals Court the highest civil servants tribunal, it is also the highest court with regard to social security. Of all tribunals the Central Appeals Court is the one most likely to be faced with reliance on the Covenant. For that reason the case law of this highest tribunal deserves special attention.

In the same ruling in which the violation of Article 7 of the Covenant was 'undone', the Central Appeals Court held that the direct applicability of any provision of the Covenant constituted an exception from the Covenant's general character. In other words, the main rule is formed by the absence of direct applicability. With this ruling the Central Appeals Court also responded to a ruling that it delivered three years earlier, in 1986. In 1986 the Central Appeals Court seemed to create the possibility of direct applicability by ruling that it need not be excluded in advance that the right to equal remuneration by its very nature would be suitable for direct applicability.[43] In 1989 the Central Appeals Court backed off. In 1991 direct applicability of the right to equal remuneration was dismissed definitively.[44]

In 1995 the Central Appeals Court took a new route, primarily with regard to Article 11 of the Covenant and the right to subsistence. The Central Appeals Court did not state whether or not Article 11 had direct applicability and established that the right compelled the government to guarantee an adequate minimum. The Appeals Court then assessed whether the cutback measure had an adverse effect on that minimum. Because no such adverse effect occurred and Article 11 had therefore not been violated, the direct applicability issue need not be considered.[45]

[40] F. Vlemminx, Een nieuw profiel van de grondrechten, The Hague, 2002.
[41] Civil Servants Tribunal Amsterdam, 12 March 1984, NJCM-Bulletin 1984, page 245 et seq.
[42] Central Appeals Court, 16 February 1989, AB 1989, 164. This contribution does not cite all court rulings denying the direct applicability of the Covenant. The picture evoked by the rulings cited is sad enough as it is.
[43] Central Appeals Court, 3 July 1986, AB 1987, 299.
[44] Central Appeals Court, 17 December 1991, RSV 1992, 164.
[45] Central Appeals Court, 31 March 1995, JB 1995, 161.

Various later rulings, concerning not only issues regarding the Covenant[46] but also issues regarding the European Social Charter and ILO Conventions, confirmed this new approach. This again opened up the possibility of the Covenant's direct applicability.

Unfortunately, the Central Appeals Court backed off for a second time. On 25 May 2004[47] the Central Appeals Court argued, further to reliance on Article 11 of the Covenant, among other things, that given the wording and the purport this article contains generally formulated objectives rather than a right that can be invoked by citizens. With reference to the explanatory memorandum on the approval act of the Covenant direct applicability was denied. The same happened on 18 June 2004 with regard to Article 9 of the Covenant.[48] Therefore, at present the Central Appeals Court is back to square one as far as the direct applicability of the Covenant is concerned.

3.4 Case law of the civil courts

In addition to the Central Appeals Court's approach, the civil court's views should be discussed as well. It appears that the Central Appeals Court's backing off was caused in part by the persistently dismissive strategy of the civil courts. When in 1990 the Central Appeals Court definitively dismissed direct applicability of the right to equal remuneration contained in Article 7 of the Covenant, it did so with reference to a similar negative ruling of the highest civil court, the Supreme Court.[49] A similar picture emerges for Article 11 of the Covenant. In 1995 and 1997 the Central Appeals Court held that the direct applicability need not be considered and that Article 11 compelled the state to guarantee a minimum. On 6 September 2000 the Court of The Hague held that Article 11 was worded so generally that no minimum could be distilled and that the Article therefore did not have direct applicability.[50] As mentioned earlier the Central Appeals Court in recent case law abandoned the approach of 1995 and 1997.

It is clear that the civil courts have never been enamoured of the Covenant's direct applicability, not even if a freedom of citizens is undeniably at stake and for that reason non-interference is expected of the state. For instance, in 1983 the right to strike contained in Article 8 of the Covenant was denied direct applicability because the Article, given the words 'The States Parties to the present Covenant undertake to ensure', does not address citizens directly, but the state.[51] The civil courts also strongly stress the importance of the government's position in the explanatory

[46] Central Appeals Court, 22 April 1997, JB 1997, 158.
[47] Central Appeals Court, 25 May 2004, LJN: AP0561.
[48] Central Appeals Court, 8 June 2004, LJN: AP4680.
[49] Supreme Court, 20 April 1990, AB 1990, 338.
[50] District Court of The Hague 6 September 2000, RAwb 2001, 55.
[51] Supreme Court, 6 December 1983, NJ 1984, 557.

memorandum on the act adopting the Covenant. It was partly because of this explanatory memorandum that the Supreme Court in 1989 did not award direct applicability to Article 13 of the Covenant,[52] explicitly stating a year later that some meaning should definitely be attached to the explanatory memorandum.[53]

It is also worth mentioning that in 1993 the Supreme Court determined that the right to equal remuneration contained in Article 7 of the Covenant could be awarded direct applicability to some extent if it were combined with Article 26 of the International Covenant on Civil and Political Rights.[54] In that dispute the Supreme Court ruled that the discrepancy in remuneration between married and unmarried persons was incompatible with Article 7 of the Covenant and Article 26 of the International Covenant on Civil and Political Rights. The direct applicability of Article 7 of the Covenant through Article 26 of the International Covenant on Civil and Political Rights appears to be practised consistently.[55] On 30 January 2004 the Supreme Court placed great value on the combination of Article 7 of the Covenant and Article 26 of the International Covenant on Civil and Political Rights but at the same time, referring to the ruling from 1990, repeated that Article 7 of the Covenant by itself does not have direct applicability.[56]

3.5 The attitude of the Dutch Government

With regard to the reporting duty imposed on the States Parties by Articles 16 and 17 of the Covenant, it should be noted first of all that the last report of the Netherlands was submitted in 1996.[57] As reports should be filed every five years, and the next report therefore should have been submitted in 2001, it appears that the obligations under the Covenant are not being taken very seriously. In the 1996 report the government's comment on the right to food is brief:

> ***'Adequate food***
>
> *235. Food production in the Netherlands greatly exceeds the population's requirements. During recent decades food production and agricultural production in general have increased rapidly and the production and import/export statistics show that the Netherlands is one of the countries*

[52] Supreme Court, 14 April 1989, AB 1989, 207.
[53] Supreme Court, 20 April 1990, NJ 1992, 636.
[54] Supreme Court, 7 May 1993, NJ 1995, 259. More than ten years earlier the administrative court had already introduced this approach, but with regard to Article 11 of the Covenant and Article 26 of the International Covenant on Civil and Political Rights; see, for instance, summary proceedings judge of Arnhem, 19 October 1978, tB/S III, 97.
[55] On the convergence of civil and political rights with social, cultural and economic rights, see the chapter by Vlemminx elsewhere in this book.
[56] Supreme Court, 30 January 2004, LJN: AM2312.
[57] Second periodic report, 05/08/96, E/1990/6/Add. 11. Since this line was written a subsequent report has appeared. That report is discussed by Coomans elsewhere in this book.

with a large net export of food. There are, however, substantial imports of food and cattle feed from various industrial and developing countries.

236. Product quality (e.g. in terms of hygiene, residue levels, contaminants and nutritional quality) is high and meets legal as well as other consumer requirements.

237. The prices of common foodstuffs are relatively low. Low-income groups are able to buy adequate amounts of food. In periods of severe recession extra attention is paid to the low-income groups. Specific measures may be taken. At the moment there are no programmes to secure the food supply of special groups. An important consideration in this respect is the percentage of income that is spent on food. This decreased from 37 per cent in 1960 to 14.9 per cent in 1993. There are no regional differences which would be relevant here.

238. The foregoing shows that food and nutrition policy in the Netherlands focuses on food quality and the promotion of healthy eating habits by the various population groups, rather than on the supply of food as such. General policy is described in the Food and Nutrition Policy report presented to Parliament in 1984, which was followed by progress reports in 1989 and 1993.[58]

239. Nutrition surveys covering 6,000 people were made in 1987 and 1992. The next one will most probably take place in 1997. The figures show an adequate supply of micro and most macronutrients. However, from the health point of view, fat consumption and to a lesser degree energy intake are considered to be too high.

240. In 1986 a long-term campaign aimed at reducing fat consumption was started. A positive downward trend (fat represented 40 per cent of total calorie consumption in 1987 and 38 per cent in 1992) has been noted. The campaign will continue in the next few years, and will include both a direct and an indirect approach to lowering fat consumption till the optimum of 35 per cent is reached.

241. Nutritional education is an important instrument of food and nutrition policy in the Netherlands. The leading organization in this field is the Food and Nutrition Information Office. This organization is mainly financed by the Government, that is to say the Ministries of Health, Welfare and

[58] For the latest report ('Our food, our heath' 2004) see: www.rivm.nl/bibliotheek/rapporten/270555009.html.

Sport and the Ministry of Agriculture. The Office directs its activities at the public, educational and information institutions and industry.

242. Environmental considerations are of growing importance in food production and consumer choice. Important topics are the use of pesticides and a more balanced use of fertilizers. Attention is particularly focused on organic farming, both nationally and by the EU. The area under cultivation and the number of organic farmers are slowly increasing. High prices, due to lower production levels and distribution problems, prevent organic farming from expanding any faster.'

When the report was considered in 1998[59] Mr Ahmed of the Committee on Economic, Social and Cultural Rights remarked:

'that, according to the Netherlands section of the International Commission of Jurists, 240,000 households, or almost 1 million persons, were living on an income below the social minimum and some 250,000 children belonging to poor families participated very rarely in recreational and cultural activities. He was surprised that a country as wealthy as the Netherlands was unable to solve those problems.'

To this Mr Potman of the Dutch delegation replied:

'There seems to be no basis for the allegation that 250,000 children were unable to exercise the rights provided for in the Covenant.'

That this figure of 250,000 children was taken from the second annual report on poverty issues,[60] which had been presented to the Minister of Social Affairs and Employment in 1997, was not mentioned. It should further be noted that three years later the official study by the Social and Cultural Planning Office and the Central Bureau for Statistics revealed that one out of seven households in the Netherlands were living below the poverty line.[61] To all appearances the Dutch government, or the delegation representing the government, has deliberately provided the Committee on Economic, Social and Cultural Rights with incorrect information. The same seems to apply to the direct applicability issue. When the report was considered in 1998 members of the Committee:[62]

---

[59] E/C.12/1998/S.R. 14.

[60] G. Engbersen, J.C. Vrooman and E. Snel (ed.), De kwetsbaren, Tweede jaarrapport armoede en sociale uitsluiting, Amsterdam 1997, pages 9-10, 130, 147 and 154.

[61] Social and Cultural Planning Office and Central Bureau for Statistics, Armoedemonitor 2000, The Hague 2000, p. 16-17. On poverty in the Netherlands see www.armoedemonitor.nl.

[62] DPI-Press Releases, Committee on Economic, Social and Cultural Rights starts consideration of report of the Netherlands, 5 May 1998.

'said they knew Dutch courts incorporated international law references in court decisions, in particular in the field of human rights, and that was admirable. Had there been any recent court decisions in which the Covenant had been mentioned?'

The Dutch delegation replied:

'The direct application of the Covenant was not needed in the Netherlands as such because whoever had problems with one of the rights enshrined therein could refer to the other instruments the Government had ratified, including International Labour Organisation conventions and the European human rights convention.'

Where the ILO conventions are concerned, this answer is also misleading. The courts almost always deny direct applicability, as with the Covenant. But there have been exceptions. In 1996 the Central Appeals Court awarded direct applicability to a provision of the ILO Convention 102 and established that a violation had occurred.[63] The government then terminated a similar European convention on social security (ILO Convention 102 could not be terminated because the applicable term had not yet arrived).[64] Fortunately Parliament refused to approve the termination.[65] This affair occurred shortly before the delegation gave the above response. The delegation knew or should have known that the government did not appreciate and would not accept the establishing of a violation of an ILO Convention, but did not mention this. On 14 March 2003 the Central Appeals Court further awarded direct applicability to a provision contained in ILO Convention 118 and again a violation was established.[66] Again the government wanted to terminate the convention and a bill to approve the termination was submitted to parliament. This time Parliament conceded. The termination of ILO Convention 118 was effected retroactively as of 14 March 2003![67]

The manner of reporting, like the case law of the courts on the Covenant, clearly shows that every effort is being made to minimise the significance of this convention for the Netherlands.

3.6 The right to be free from hunger in Dutch foreign policy

Dutch foreign policy seems to show two different faces. On the one hand positions are taken in support of the right to food, while on the other hand the Netherlands are not willing to enter into binding obligations.

[63] Central Appeals Court, 29 May 1996, RSV 1997, 9.
[64] Parliamentary Documents II 1996-1997, 25 524.
[65] Acts II 1997-1998, page 4381.
[66] Central Appeals Court 14 March 2003, RSV 2003, 114.
[67] Act of 9 December 2004, Staatsblad (*Statute Book*) 715.

3.6.1 Human rights and development co-operation policy

Dutch policies on human rights and development co-operation touch on the right to food. Human rights policy is a part of development co-operation policy. The latter focuses on poverty reduction. In this respect the Netherlands take the Millennium Development Goals (MDGs) as a guideline. The Netherlands are convinced that eradication of poverty will contribute to the eradication of hunger, which is seen as an indicator of poverty.

3.6.2 Bilateral co-operation

In bilateral development relations, the Netherlands opt for 'unbound' food aid. This is to say that, as far as possible, food is purchased locally. This approach is intended to stimulate the local economy and the local possibilities to generate an income. In the Dutch view food shortages are not usually caused by an absolute lack of food, but by a lack of means to purchase the food that is available.

3.6.3 Multilateral level

At the multilateral level, the Netherlands support initiatives to enhance food security by WHO, UNHCR and FAO. Here also the MDGs are leading.

3.6.4 FAO voluntary guidelines

Against the backdrop of the foregoing, one would expect staunch Dutch support for any initiative to promote the realisation of the right to food. Such expectations meet with disappointment.

With a view to implementing the Plan of Action of the 1996 World Food Summit, in 2002 the Council of FAO instituted an 'Intergovernmental Working Group For The Elaboration Of A Set Of Voluntary Guidelines To Support The Progressive Realization Of The Right To Adequate Food In The Context Of National Food Security' (IGWG). Some 200 representatives from 90 FAO Member States took part in the four IGWG meetings that took place between March 2003 and September 2004. At the 4th meeting the Council of the FAO adopted the guidelines.[68] 'The objective of these Voluntary Guidelines is to provide practical guidance to States in their implementation of the progressive realization of the right to adequate food in the context of national food security'.[69]

---

[68] See: www.fao.org/legal/rtf/rtf-e.htm.

[69] Voluntary Guidelines to support the progressive realisation of the right to adequate food in the context of national food security, § 6.

Already in 2002 the European Commission cautioned the EU Member States about the possible implications if the Guidelines would have to be implemented on the international plane. The Member States' policies on agriculture, trade and finance might not be compatible with these guidelines.[70]

The EU was not disposed to agree to binding obligations to ensure food security within its own jurisdiction, and even less an obligation to provide humanitarian aid, including food aid, to any country in need that would ask for it.[71] At the time of the negotiations, the Netherlands held the EU presidency.

The Netherlands instructed its negotiators to take a 'constructive' approach in order to control damage as far as possible.

Four positions were taken:
1. human rights conventions give the member states obligations vis-à-vis their own population;
2. voluntary guidelines are not to be considered as legally binding documents, but practical tools to help realise the right to adequate food at the national level;
3. the voluntary guidelines do not bind;
4. the voluntary guidelines do not address 'obligations' of states in international relations.

Despite opposition from developing countries, all four positions were upheld.

4. Dutch food law outside the context of ICESCR

4.1 EU/Dutch food safety law

We have seen above that Article 11 has little or no impact on the Dutch legal system. The government presumed that the right to adequate food had already been fulfilled in national law.

In interpretative documents to Article 11 ICESCR like General Comment nr. 12[72] of the UN Economic and Social Council and the FAO Voluntary Guidelines to support the progressive realization of the right to adequate food in the context of national food security,[73] the concept of adequate food has been specified to encompass: 'the availability of food in a quantity and quality sufficient to satisfy the dietary needs of individuals, free from adverse substances, and acceptable

[70] EC, Issues Paper, The Right to Adequate Food, 16 December 2002.
[71] Neither did the USA.
[72] E/C.12/1999/5, 12 May 1999. Available at: www.unhchr.ch/tbs/doc.nsf/0/3d02758c707031d58025677f003b73b9?Opendocument.
[73] Available at: www.fao.org/docrep/meeting/009/y9825e/y9825e00.htm.

within a given culture'. In other words: adequacy not only concerns nutrition but also safety and acceptability. In the Netherlands and, more in general in the EU, extensive regulatory systems deal with the issues of social security, of food safety and to a lesser extent of cultural acceptability. These are, however, not recognised as human rights issues. No connection has been made to Article 11 ICESCR.

We now turn to these other provisions, in particular those on food.

Dutch legislation does not recognise an enforceable right of access to food as such. However there exists a large body of social security legislation that ensures a minimal financial standard of living in most cases.

Furthermore, the Netherlands has a body of food law consisting of EU regulations and national legislation modelled on EU directives, that addresses questions of food quality - especially food safety - and nutrition.

4.2 Framework legislation

Dutch food law can be characterised as economic regulatory law. The current legislative situation dates back to 1919 when the first national Commodities Act[74] (*Warenwet*) was published. This act is a framework act that contains the legal basis for statutory legislation in the form of Orders in Council[75] and Ministerial Orders[76] on both food and non-food.[77] Before 1919 provisions of food law were issued at municipal level. At present, the 1919 Act is still used, but it has been considerably amended and expanded. The major objectives of the act are the protection of public health, safety, fair trade and proper information about products.[78]

To a large extent, statutory legislation implements European directives and is thus harmonised with food legislation elsewhere in the European Union. Since the publication of the 2000 White Paper on Food Safety,[79] European food law is under review. The core of modern European food law no longer consists of directives, but instead the regulation is chosen as the most important legislative instrument. As a consequence, national food law is increasingly being replaced by directly applicable European legislation. First amongst these regulations is Regulation 178/2002, the so-called General Food Law ('GFL').

[74] Act of 28 December 1919, Staatsblad (*Statute Book*) 1935, 793. Sometimes translated as: Food and Commodities act.
[75] Usually referred to as: 'Warenwetbesluiten'.
[76] Usually referred to as: 'Warenwetregelingen'.
[77] Such as toys and cosmetics.
[78] Article 3 Commodities Act.
[79] COM(1999) 719 final.

4.3 The system of food law

For the sake of analysis the heterogeneous body of food legislation can be categorised as follows.

First a distinction can be made between substantive and procedural food law.

Substantive law focuses on the content of legal relations. Procedural law focuses on how to realise this content. These include procedures concerning pre-market approval of food products, inspection, monitoring and enforcement - including provisions that attribute the necessary powers to public authorities - and legal protection against measures of the authorities.

It appears that most of the substantive rules and regulations can be categorised in a limited number of ways. Three[80] types are easily distinguished: (1) rules concerning the substance of food, (2) rules concerning the handling of food, and (3) rules concerning communication about food.

The distinction between these three types is not watertight. One can often argue as to whether a certain rule is of one type or another[81] but that does not diminish the usefulness of the distinction.

4.4 The substance of food

4.4.1 Food safety

Generally speaking, food producers are free to choose their ingredients.[82] Exceptions apply to specifically designated categories such as food additives, genetically modified foods and (other) novel foods. These need to pass a safety assessment before they can be brought to the market.

Besides raw materials that the producer intentionally includes in a food product, many kinds of chemicals or (micro) organisms may unintentionally find their way into the final product before it reaches the consumer.

---

[80] In its definition of food law, German literature distinguishes some more different types: extraction, production, composition, quality, labelling, packaging and designation.
[81] For instance, the rule that a product must comprise at least 35% cocoa in order to be called chocolate can be considered a rule on content (35%) or a rule on communication, 'if you want to call it chocolate...'.
[82] However, if they want to use certain protected names of designations they may have to comply with vertical compositional standards legislation. In the Netherlands there are, for instance, Warenwetbesluiten giving standards for the composition of honey, chocolate, jam and bread.

This situation is covered to a certain extent by the general rules on food safety, which insist that no dangers to the consumer be present.[83] However, there are also more specific rules. First amongst these is the Framework regulation (EC) no. 315/93. This Regulation opens the possibility to establish maximum content levels for certain substances in food. If these levels are surpassed, the food may not be brought to the market. 'Contaminant' is defined as:

> *'any substance not intentionally added to food which is present in such food as a result of the production (including operations carried out in crop husbandry, animal husbandry and veterinary medicine), manufacture, processing, preparation, treatment, packing, packaging, transport or holding of such food, or as a result of environmental contamination. Extraneous matter, such as, for example, insect fragments, animal hair, etc, is not covered by this definition.'*

Regulation (EC) no. 466/2001 gives maximum levels for a whole range of contaminants, for instance for nitrates in lettuce and spinach, and aflatoxins in nuts, dried fruit, grain, herbs and milk.

A particular kind of contaminant is residue from veterinary medicine, or from additives in feeding stuffs (in animal products like meat, milk or eggs), or crop protection products (in plants). For these products maximum residue levels (MRLs) can be established. If these levels are surpassed, the food may not be brought to the market.

4.4.2 Nutritional quality

At present no mandatory requirements exist with regard to the nutritional quality of food. In the Netherlands, government activity in this field is restricted to providing the public with information through the Food and Nutrition Information Office *(Voedingscentrum)*[84].

There is no obligation for food business operators to comply with nutritional quality standards, nor to give nutrition information on the label.[85] Some rules do exist, however, in case such information is provided voluntarily.[86]

[83] Article 14 GFL.

[84] www.voedingscentrum.nl.

[85] However, a proposal is pending to introduce such an obligation (COM(2008) 40).

[86] Directive 90/496/EC; Warenwetbesluit Voedingswaarde-informatie levensmiddelen. A proposal for mandatory nutrition labelling in the EU is pending; COM(2008) 40 final.

4.5 Food handling

According to Article 17 of the General Food Law, food business operators are responsible for meeting the relevant (food safety) requirements at all stages of production, processing and distribution of food.

There are several kinds of rules addressing the handling of food. There are those that aim to prevent food safety problems from occurring, those that aim to be prepared if food safety problems might occur and those that give obligations and instruments to deal with food safety problems when they occur.

4.5.1 Preventing problems – Hygiene

The safety of food products when they reach a consumer's plate depends largely on the way they have been produced. For this reason rules have been set to ensure that safe methods of production are used.

Regulation (EC) No. 852/2004 of the European Parliament and of the Council of 29 April 2004 on the hygiene of foodstuffs,[87] was published in the Official Journal of the European Union on 30 April 2004. It entered into force on 1 January 2006, replacing existing national legislation in the Member States.[88]

The word 'hygiene' is taken in a broad sense. It means measures and conditions necessary to control hazards and ensure the fitness for human consumption of a foodstuff taking into account its intended use.

The Regulation imposes a general obligation on food business operators to ensure that in all stages of production, processing and distribution food under their control satisfies the relevant hygiene requirements.

At the heart of these requirements is the so-called HACCP (hazard analysis and critical control points) system. This is a system of (imposed) self-regulation in which the food business operator must analyse the processes in his or her business in order to identify the hazards that may occur. Then s/he has to identify the critical control points and the limits at those points that separate acceptability from unacceptability. The operator must implement monitoring procedures at the critical control points and prepare for corrective actions if a control point is not under control. Finally, everything that is undertaken must be documented and recorded and the information must be made available to the public authorities.

---

[87] And Regulation (EC) No 853/2004 of the European Parliament and of the Council of 29 April 2004 laying down specific hygiene rules for food of animal origin.
[88] In the Netherlands, for instance, the 'Warenwetbesluit Bereiding en behandeling van levensmiddelen', Staatsblad (*Statute Book*) 1992, 678; and 'Warenwetregeling Hygiëne van levensmiddelen', Staatscourant 1994, 245. All Dutch legislation – in Dutch language – can be found at www.overheid.nl.

The application of the HACCP principles can be facilitated and encouraged by national or Community guides for good practices.

In its annexes the Regulation gives general hygiene requirements for primary production and for all food business operators. Most of these requirements are concerned with cleanliness and prevention of cross contamination.

4.5.2 Preparing for problems – Traceability

The General Food Law[89] requires that food and food ingredients be traceable. This means that Food business operators must be able to identify (upstream) any person from whom they have been supplied with a food, a feed, a food-producing animal, or any substance intended to be, or expected to be, incorporated into a food or feed; and (downstream) to identify the other businesses to which their products have been supplied.

This information shall be made available to the competent authorities on demand.

The intention of the traceability system is to enable food safety problems to be identified at the source, and across the food chain.

4.5.3 Dealing with problems – Withdrawal and recall

Food business operators may not bring food to the market if it is unsafe.[90] If unsafe food nonetheless is discovered to have made it to market, then the product must be withdrawn from downstream businesses or recalled from the consumer.[91]

The food business operator, who considers or has reason to believe that a food it has imported, produced, processed, manufactured or distributed, is not in compliance with the food safety requirements, has at least four duties.

First, there is the duty to immediately initiate procedures to withdraw the food in question from the market.

Second, the operator must immediately inform the authorities that s/he has reason to believe that an unsafe food has been placed on the market. The operator must also inform the authorities of all actions taken to deal with the problem.

Third, in case the product may have already reached consumers, the operator shall effectively and accurately inform those consumers of the reason for its

[89] Article 18 GFL.
[90] Article 14 GFL.
[91] Article 19 GFL.

withdrawal, and recall products already supplied when other measures are deemed insufficient to achieve a high level of health protection.

Fourth, the food business operator has a duty to collaborate with the competent authorities on actions taken to avoid or reduce risks posed by foods, which they supply.

4.6 Labelling

An important issue in consumer protection is to ensure that the consumer is aware of what s/he consumes. Many rules exist concerning the obligation of food business operators to provide the consumer with adequate information by labelling.

The most important codification of these rules is to be found in Directive 2000/13 of the European Parliament and of the Council of 20 March 2000 on the approximation of the laws of the Member States relating to the labelling, presentation and advertising of foodstuffs: the so-called 'Labelling directive'.[92] This directive has been implemented in the Netherlands in the 'Warenwetbesluit Etikettering van levensmiddelen'.

Labelling means 'any words, particulars, trade marks, brand name, pictorial matter or symbol relating to a foodstuff and placed on any packaging, document, notice, label, ring or collar accompanying or referring to such foodstuff'. Labelling may not be misleading.

All pre-packaged food products must be labelled in a language that is easily understood. Usually this means in the national language of the Member State. Other information is mandatory, restricted or forbidden.

There are ten required (mandatory) pieces of information:
1. the name under which the product is sold;
2. the list of ingredients;
3. the quantity of certain ingredients or categories of ingredients;
4. in the case of pre-packaged foodstuffs, the net quantity;
5. the date of minimum durability or, in the case of foodstuffs which, from the microbiological point of view, are highly perishable, the 'use by' date;
6. any special storage conditions or conditions of use;
7. the name or business name and address of the manufacturer or packager, or of a seller established within the Community.

[92] OJ 6.5.2000 L 109/29. A proposal to replace this directive by a regulation is currently in procedure (COM(2008) 40).

8. particulars of the place of origin or provenance where failure to give such particulars might mislead the consumer to a material degree as to the true origin or provenance of the foodstuff;
9. instructions for use when it would be impossible to make appropriate use of the foodstuff in the absence of such instructions;
10. with respect to beverages containing more than 1.2% by volume of alcohol, the actual alcoholic strength by volume.

Specific labelling requirements demand that the presence of allergens, additives, novel ingredients and GMOs be mentioned on the label.

These labelling requirements are rather extensive. In most cases information that is relevant from the point of view of acceptability of the food product for certain cultural groups in society, will therefore be available.

4.7 Legal protection

Individuals have several opportunities to bring proceedings contesting infringements of their rights under food law to court.

4.7.1 Judicial review

Interested parties can challenge all administrative decisions by administrative bodies on the basis of the General Administrative Law Act *(Algemene wet bestuursrecht)*.[93] If, for instance, subsistence under social security legislation is denied, this decision can be contested first by lodging an objection with the body that took the decision and after that by lodging an appeal with the competent administrative court.

4.7.2 Tort law

Individuals can bring any case to the civil courts if they claim that they suffer damages from someone else's unlawful behaviour.[94] It is irrelevant whether this behaviour has been committed by private parties or public authorities.[95]

Behaviour is considered unlawful if the perpetrator neglects his legal obligations, infringes on the victim's legal rights or does not exercise sufficient diligence with regard to the victim's legitimate interests. Furthermore, the perpetrator must be at fault.

[93] For an English translation see www.justitie.nl/images/Engelse%20tekst%20Awb_tcm34-2020.pdf or www.globalcompetitionforum.org/regions/europe/Netherlands/GENERAL%20ADMINISTRATIVE%20LAW%20ACT.pdf.
[94] Article 6:162 Civil Code.
[95] However, if the interested party has recourse to a procedure under administrative law, the civil courts will not hear the grievance.

At present there does not seem to be a rule obliging a restaurant to provide meals that provide a balanced diet or to warn clients in case the meals offered are nutritionally unbalanced. It seems unlikely therefore that a lawsuit holding a restaurant responsible for obesity, like the one against McDonalds in the USA, could be won at present. The poor quality of fast food from a nutritional point of view is considered to be known to the general public. This makes it everybody's individual responsibility to vary food intake.

In the European Union, discussions are ongoing as to whether a system of mandatory nutrition labelling should be introduced in order to combat obesity. If such requirements are correctly followed, it would help the food business operator in his defence to show that the consumer was in a position to make informed choices.

If, on the other hand, the consumer suffers damages because the labelling requirements have not been complied with, s/he has an easy case. In case of financial damage – which may occur if allergens were not mentioned on the label – the victim may sue for compensation. In case of non-material damages, as will be the case when cultural information (GMO, pork) is omitted on the label, it seems more appropriate to ask for an injunction.

In the case of novel foods it is mandatory to indicate the presence of materials which give rise to ethical concerns.[96] The presence of GMOs must be labelled. In other cases it will depend on the general rules on labelling as to whether cultural concerns have to be visible on the label.

4.7.3 Product liability law

The food business operator who, by bringing unsafe food to the market causes a customer to suffer damages, can be held liable under tort law if it can be proven that he was at fault.

Besides tort law, a more specific regime concerning defective products also exists and covers food products as well.[97]

The rules on product liability have been harmonised in the European Union by Directive 85/374.[98] The Dutch civil code follows this directive to the letter. Directive 85/374 lays down the principle of strict liability of the producer, which

[96] Article 8 (1)(c) Regulation (EC) 258/97.
[97] In writing this section, use has been made of Nicole Coutrelis, Product Liability in the Food Sector, http://ec.europa.eu/enterprise/regulation/goods/docs/liability/1999-greenpaper-replies/046.pdf and of Bernd van der Meulen, Productaansprakelijkheid voor onveilige levensmiddelen, Journaal Warenwet 2004 (2), pp 3-8 and (3) pp. 3-11.
[98] As amended by Directive 1999/34.

means that a producer may be held responsible for a damage caused by a defective product s/he has put on the market even in the absence of fault.

The producer within the meaning of the directive is not only the manufacturer of the final product, but also any other person in the chain who has produced raw materials or a component of the product, or any person who, 'by putting his name, trade mark or other distinguishing feature on the product, presents himself as its producer'. All these persons bear their own liability, and consequently the victim may make a claim against any of these persons for complete compensation. The *supplier* is not responsible, except in cases when s/he does not, or cannot inform the victim of the producer's identity.

The plaintiff (victim) must prove three conditions:

1. A damage to the person (death or personal injury) or to the person's private property (damages to professional property are not covered by the directive nor the civil code). For the food sector, this implies that only the final consumer can invoke product liability.
2. A defect of the product, which is established when the product does not provide the safety that one is entitled to expect, taking into account particularly its presentation, the use which could be reasonably expected and the time the product has been in circulation. A mere lack of conformity to specifications (stemming from regulations or from a contract) is not sufficient to declare a product defective if safety is not at stake.
   In the field of food, due to the amount of food safety requirements, consumers are entitled to expect a high degree of safety. The producer can influence the expectations of the consumer by providing specific information, instructions on use and storage on the label.
3. A causal relationship between the defect and the damage.
   Consumers of unsafe food may experience some difficulty in proving the causal relationship between a specific food and the (health) damages that have been suffered. Microbiological hazards become apparent only after a certain incubation time. The food disappears by its very use and usually people use large varieties of food. This often makes it difficult to determine the cause of a foodborne damage.

In order to disclaim liability, a producer has to prove the existence of one (or several) of the following circumstances:

- s/he did not put the product in circulation;
- the product was not defective when put into circulation;
- the product was not manufactured by the producer for sale or any form of distribution;
- the defect is due to compliance with mandatory regulations issued by public authorities;

- the state of scientific and technological knowledge at the time when the product was put into circulation did not allow the existence of the defect to be discovered ('development risk');
- the defect of the final product (in case the manufacturer of a raw material or component is held liable) is attributable to the design of the final product or to the instructions of the manufacturer;
- the damage is caused, totally or in part, by the fault of the victim.

In practice few product liability cases occur in the food sector. On the one hand it is difficult for consumers to prove their case and often damages are too low to bother trying. On the other hand, food business operators do not like the bad publicity that may be involved in a court case and go to some length to keep their consumers happy by dealing with complaints in a generous way.

4.8 Influence on commerce

Only in so far as food safety aspects are concerned does the right to adequate food influence commerce. The above-mentioned General Food Law (Regulation EC/178/2002), which is directly applicable in the Netherlands, holds food business operators responsible for compliance with food safety law. The General Food Law takes a so-called holistic approach. This means that food (safety) law applies to all the players in the entire food chain ('from farm to fork').

The inclusion of primary production in the ambit of food law is a recent development. The European Union puts pressure on farmers to comply with food law by including Regulation 178/2002 in a list of regulations that can be sanctioned by loss of subsidies.[99] This is the so-called 'cross compliance'. The name indicates that there is no connection between these regulations and the subsidies, other than that the subsidy is used as a sanctioning mechanism.

The General Food Law holds the Member States responsible for the enforcement of food law. This has been further elaborated by some regulations concerning official controls.[100]

5. Conclusion

The research for writing this report did little to enhance our feelings of chauvinism. The Netherlands enjoys wealth that would, in comparison to many other countries

[99] Regulation 1782/2003.

[100] Regulation (EC) No 882/2004 of 29 April 2004 on official controls performed to ensure the verification of compliance with feed and food law, animal health and animal welfare rules, OJ L 191/1 28.5.2004; Regulation (EC) No 854/2004 of 29 April 2004 laying down specific rules for the organisation of official controls on products of animal origin intended for human consumption, OJ L 226/83 25.6.2004.

in the world, make it rather easy to develop a working system of respect for the human right to food. Nevertheless, we found that in this respect Dutch law and politics leaves much to be desired.

Dutch food law has developed outside the scope of the right to food. It is modelled on European directives and to some extent replaced by European regulations. Dutch food law is more concerned with food safety than with availability or nutritional quality of food. It does, however, protect consumers from unsafe food.

Further reading

On Dutch law in general

Leonard F.M. Besselink, Constitutional Law in the Netherlands, Ars Aequi Libri Nijmegen, 2004.

Fred J. Bruinsma, Dutch Law in Action: www.uu.nl/content/dutchlawinaction2003.pdf.

Jeroen Chorus (c.s. eds.) Introduction to Dutch Law, 4th ed. Kluwer Law International 2006.

Oswald Jansen and George Middeldorp, Researching Dutch Law www.llrx.com/features/dutch.htm).

Sanne Taekema (ed.) Understanding Dutch Law, BJu (www.bju.nl) The Hague 2004.

On right to food and food law

The European Institute for Food Law provides a useful portal to Internet sources on right to food and food law: www.food-law.nl.

Nicole Coutrelis, Product Liability in the Food Sector, http://ec.europa.eu/enterprise/regulation/goods/docs/liability/1999-greenpaper-replies/046.pdf.

B.M.J. van der Meulen, The Right to Adequate Food. Food Law Between the Market and Human Rights (inaugural address) ISBN 90 5901 9628, Elsevier 2004.

B.M.J. van der Meulen and M. van der Velde, European Food Law Handbook, Wageningen Academic Publishers, 2008 (www.wageningenacademic.com/foodlaw).

Chapter 3
Dutch schizophrenic constitutional law and enlightening practices of the European Court for Human Rights

Frank Vlemminx

1. Introduction

The right to adequate food is laid down in Article 11 of the International Covenant on Economic, Social and Cultural rights (ICESCR). Article 11 elaborates upon Article 25 of the Universal Declaration of Human Rights dating from 1948 that guarantees everyone the right to an adequate standard of living. In turn this Article 25 of the Universal Declaration was inspired by the so-called 'four freedoms' speech with which the American President Roosevelt gave extra force to the allies' battle against fascism. The sacrifices made by the allies were to open the way to the 'freedom of speech', the 'freedom of worship', the 'freedom from fear' and, the subject of our discussion here, the 'freedom from want'.

When we consider the right to adequate food our thoughts immediately go out to famine in the Third World and shocking pictures of emaciated children, which play upon the Western conscience. However, the effect of these pictures is limited and short-lived. There is a strong inclination to put famine and death by starvation down to fate, to indefinable forces such as climate and economy or even a lack of discipline or willingness to work on the part of the victims themselves. Our conscience regrets that things should be this way, but there is nothing that can be done about it. This is why it is important to remember that the right to adequate food is linked to our Second World War, our Winter of Starvation, and our post-war Reconstruction as well as to the pillars upon which we established our modern Dutch state in the second half of the previous century. In this modern state famine and death from starvation can no longer be explained away with reference to fate, economic forces or personal fault. Famine and death from starvation have been banished from the Netherlands: no one may be refused or denied adequate food. This principle point of departure finds its equivalent in Article 11 ICESCR. Discrepancies between that stipulated in Article 11 ICESCR and daily reality in the Netherlands are unacceptable. Which is surely reason for us to presume that the most basic right to adequate food embodied in Article 11 is held in extra high esteem in the modern Dutch state and is fully enforceable by the Dutch courts. No major problems need be anticipated. Few appeals will be filed. Seldom are standards, ideal and practice so much in agreement.

2. Schizophrenic constitutional law

And yet this presumption is incorrect. Despite the special fundamental character of the right to adequate food and despite the fact that there is more than enough food available in the Netherlands, demanding recognition of the right laid down in Article 11 in the Dutch court is completely pointless. Is this a paradox? Is this an apparent inconsistency that is nevertheless based on sound legal arguments? Or is it a serious contradiction, an example of schizophrenic constitutional law? I fear the latter is true. The explanation of this remarkable state of affairs is highly legally technical, complex in the extreme, and generally lacking in conviction. It consists of two separate yet interacting parts. On the one hand there is friction because the right to adequate food is a so-called social fundamental right and not a classic fundamental right such as the freedom of expression or the freedom to worship. On the other hand there is friction because Articles 93 and 94 of the Dutch Constitution severely restrict the possibility of relying upon rights derived under international treaties in the Dutch courts.

It is generally accepted that since the early twentieth century the so-called monist system applies in the Netherlands with regard to international law. This means that international rules of law are automatically integrated into the national legal system.[101] The right to adequate food therefore forms an integral part of the Dutch legal order. The question then arises as to whether it is possible to test the international legal rule. There are indeed rules that may impose obligations upon the national authorities, but at the same time these rules leave the national authorities with so much room for interpretation that it is impossible or virtually impossible for the courts to assess whether the rules are being correctly applied. This is the problem of direct applicability. If the rule lends itself to assessment by the courts, it is directly applicable. This issue is also the subject of discussion in other countries and the phenomenon is referred to by various names such as 'self-executing', and 'direct legal effect'. In the mid-twentieth century the Netherlands embodied its own variation of direct applicability in Articles 93 and 94 of the Dutch Constitution. A treaty provision is directly applicable (self-executing) if this is 'binding on all persons' within the meaning of these articles. Three other factors play a role besides the question of whether the treaty provisions leave the national authorities plenty of room for interpretation or no room at all.[102] Firstly, whether the phrasing of the treaty provision directly addresses the citizen or the state is taken into account. Secondly, the vision of the government carries great weight. In the explanatory memorandum to the bill approving the treaty the government always indicates whether or not it believes that the provisions

---

[101] This is determined by the court in case law such as HR 3 March 1919, *NJ* 1919, p. 371 (Grenstraktaat Aken).

[102] See in detail hereover: F.M.C. Vlemminx and M.G. Boekhorst, Articles 93 and 94, in: A.K. Koekkoek, De Grondwet, Een systematisch en artikelsgewijs commentaar, Deventer 2000, p. 455-478.

can be deemed to be 'binding on all persons'. Thirdly, in general the Dutch courts are inclined to approach the question of whether a provision is 'binding on all persons' in a very abstract way; they base their opinions purely on the wording of the provision without involving the issue being dealt with in the case at hand. Time and time again Article 11 ICESCR is denied a self-executing status because of these three factors, and in particular the latter two. They capitalize on the fact that the right to adequate food is a social human right.

In the first place the phrasing of the international social fundamental rights generally addresses the state and does not address the citizen directly. All the rights contained in the ICESCR, including the right to adequate food in Article 11, open with the words 'The States Parties to the present Covenant recognize the right of...' or a variation thereof. In the past this phrasing has been used as grounds for denying the right to strike as laid down in Article 8 ICESCR self-executing status. [103]

Secondly, the Dutch government argues that not only do the social fundamental rights in the Dutch Constitution leave the national authorities with so much room for interpretation that there is no room for the intermission of the courts, but that this can also be said of the equivalent rights in the treaties. In the explanatory memorandum to the Act of Approval of the ICESCR the government remarks that the rights contained in this Covenant are generally not self-executing.[104]

Thirdly, the courts' approach is very much in abstracto, or, in other words very detached from the rights contained in the ICESCR. Criteria such as 'adequate' in, for example, 'adequate food' in Article 11 come across as loose terms and it is left to the legislator or the government to determine the exact definition of 'adequate'.

There are numerous court rulings in respect of Article 11 ICESCR, but without exception these all slot into the sad framework outlined above. For this reason the discussion is limited to 3 examples. In 2005[105] the highest social security court in the Netherlands, the Central Court of Appeal, refused to regard Article 11 ICESCR as 'binding on all persons' because the article contains only indefinite objectives and because in the explanatory memorandum the government also rejects the idea that the provisions are self-executing. In 2007[106] the Administrative Judiciary Division of the Council of State adopted exactly the same position. The civil court too refuses to regard Article 11 ICESR as 'binding on all persons'. A judgement passed by the district court at The Hague 7 years ago is worthy of a special mention because it is a striking illustration of this highly abstract

[103] HR 6 December 1983, NJ 1984, 557.
[104] TK 1975-1976, 13 932 (R 1037), nr. 3, p. 12-13.
[105] CRvB 11 November 2005, LJN: AU5600.
[106] ABRvS 19 April 2007, LJN: BA4289.

approach by the courts. In this case asylum seekers who were denied any form of relief relied upon the right to adequate food. These were people without work or money of their own and it was thus clear to everyone that the absolute refusal to provide food was contrary to Article 11 ICESCR. Yet still the court refused to regard Article 11 as being 'binding on all persons' because in criteria such as 'adequate food' the phrasing of the article was too general. This was all the more embarrassing because at the same time the court recognized the importance of it being possible to appeal against violations of the right contained in Article 11 ICESCR in court. In plain language the ruling boils down to the following. If you're starving and need bread, the court itself says you can ask the court for bread. After all, Article 11 says you're entitled to enough bread. But because the court doesn't know exactly what enough bread amounts to, it gives you nothing. This is schizophrenic constitutional law at its best.

The dismissive attitude of the courts discussed above is not restricted to the ICESCR. It affects all international fundamental social rights, including those embodied in, for instance, the European Social Charter (ESC). Thus in 2003[107] the Central Court of Appeal refused to deem the right to social security contained in Article 12 ESC as 'binding on all persons' with reference to the government's remarks in the explanatory memorandum to the approval act. We will return to Article 12 ESC in paragraph 5. This case law of the Dutch courts on the national and international fundamental social rights was once dubbed 'Madurodam case law' (Madurodam is a miniature city in The Hague).[108] Here the Netherlands presents itself as miniature indeed!

3. International criticism

The refusal of the courts to control compliance with the rights contained in the ICESCR is not a matter of course everywhere. Thus the courts in Belgium see many more opportunities than do their counterparts in the Netherlands.[109]At an international level the Dutch attitude also attracts criticism. In 2006 the 'Committee on Economic, Social and Cultural Rights' that controls compliance with the ICESCR via an assessment of country reports ordered the Netherlands:[110]

> *'to reassess the extent to which the provisions of the Covenant (ICESCR) might be considered to be directly applicable. It urges the State Party to*

[107] CRvB 5 December 2003, LJN: AO2554.
[108] F.M.C. Vlemminx, Onze tandeloze sociale grondrechten in het licht van twintig jaar EVRM, NJCM-Bulletin 2003, p. 234.
[109] See: F.M.C. Vlemminx, De autonome rechtstreekse werking van het EVRM; De Belgische en Nederlandse rechtspraak over verzekeringsplichten ingevolge het EVRM, preadvies voor de NVVR, Deventer 2002, p. 28-41.
[110] UN Committee on Economic, Social and Cultural Rights, Concluding Observations on the Netherlands (24 November 2006), UN Doc. E/C.12/NLD/CO/3, par. 19.

> *ensure that the provisions of the Covenant are given effect by its domestic courts as defined in the Committee's General Comment 3, and that it promotes the use of the Covenant as a domestic source of law.'*

This 'rap over the knuckles' was long in coming. Eight years earlier, in 1988, this committee had already devoted a general comment to the necessity and the possibility of giving legal effect to the Covenant's provisions within the legal systems of the member states.[111] This comment referred to an earlier comment dating from 1990 which gave examples of ICESCR provisions that were undoubtedly directly applicable by the national courts.[112] In a nutshell the committee has been calling for more action on the part of the domestic courts for more than 15 years, in the Netherlands' case, wholly in vain. The committee is not a court. The arguments and conclusions of this body can exert moral pressure, but they are however, not legally binding. For this reason the Netherlands can set them aside without fear of punishment.

4. The ECHR and the ECtHR

International supervision of the treaties on social human rights is vested exclusively with bodies having no legal authority. In addition to the ICESCR committee one may consider the European Committee for Social Rights that monitors compliance with the ESC, and the Committees of experts that operate within the scope of the ILO conventions. International supervision of the treaties on classic human rights is also in the hands of similar bodies but with one major exception: compliance with the European Convention for the protection of Human Rights (ECHR) is supervised by the European Court of Human Rights (ECtHR) in Strasbourg. The ECtHR is a supranational[113] court before which individuals can file complaints, and which passes legally binding rulings on grounds of Article 46 ECHR. Although it is true that the ECHR contains classic rights and not social rights, the ECtHR takes the line that the boundary between the classic and the social rights is not watertight.[114] Among other things this means that the right to property guaranteed under Article 1 of Protocol No.1 is held applicable to benefits[115] and that this applicability is not dependent upon whether or not these benefits are contributory.[116] Therefore citizens can endeavour to enforce compliance with the fundamental social rights in the ESC, ICESCR and ILO conventions via an appeal to the ECHR. Two issues are important here. First the phenomenon of the convergence of human rights

---

[111] General Comment 9, 3/12/98, The domestic application of the Covenant.
[112] General Comment 3, 14/12/90, The nature of States parties obligations.
[113] ECtHR 6 February 2003, Appl.nos. 46827/99 and 46951/99, (Mamatkulov and Abdurasulovic), confirmed by ECtHR (Grand Chamber) 4 February 2005.
[114] ECtHR 6 July 2005 (dec), Appl. nos. 65731 & 65900 (Stec and Others); see also ECtHR 9 October 1979, Appl.no. 6289/73 (Airey).
[115] ECtHR 16 September 1996, Appl.no. 17371/90 (Gaygusuz).
[116] ECtHR 6 July 2005 (dec), Appl. nos. 65731 & 65900 (Stec and Others).

that is increasingly being endorsed under the ECHR and secondly the possibility to bring direct action against legislation and policy under the ECHR. Both subjects will be reviewed briefly below.

5. Convergence of human rights under the ECHR

As stated earlier the ECtHR is a supranational court that passes rulings with reference to the ECHR. The supranational status means that the ECtHR can take into account any other conventions concluded by the member state in question, irrespective of whether or not this member state deems such a convention or a specific provision in the convention to be directly applicable.

A fine example can be found in Article 3, paragraph 1 of the Convention on the Rights of the Child which stipulates that the interests of the child should always be the first consideration. In the Netherlands this article is not regarded as 'binding on all persons' insofar as it regards jurisdiction concerning aliens.[117] But in the Mubilanzila Mayeka case[118] involving a 5-year-old Congolese girl the ECtHR linked Article 8 ECHR to Article 3, paragraph 1 of the Convention of the Rights of the Child:

> *'In this connection, in the absence of any risk of the (...) applicant's seeking to evade the supervision of the Belgian authorities, her detention in a closed centre for adults was unnecessary. Other measures could have been taken that would have been more conducive to the higher interest of the child guaranteed by Article 3 of the Convention on the Rights of the Child.'*

In the same case the ECtHR indeed refers to international obligations in very general terms:

> *'In this connection, detention in centres used for aliens awaiting deportation will be acceptable only where it is intended to enable the States to combat illegal immigration while at the same time complying with their international obligations, including those arising under the Convention on the Rights of the Child signed in New York in 1989 (and by Belgium in 1991).'*

It is no surprise that where necessary, within the scope of the ECHR, the ECtHR also refers to the ESC and the conclusions of the European Committee for Social Rights. Indeed the ECHR and the ESC were both concluded within the framework of the Council of Europe. If the European Committee for Social Rights concludes

[117] ABRvS 23 September 2004, LJN: AR3385; ABRvS 22 February 2006, JV 2006, 132; ABRvS 20 March 2006, JV 2006, 165.
[118] ECtHR12 October 2006, Appl.no. 13178/03 (Mubilanzila Mayeka).

that an act is contrary to Article 5 of the ESC, which concerns the freedom to join a trade Union, it is self-evident that the act is also contrary to the freedom of assembly in Article 11 ECHR. In the Sørensen & Rasmussen case of 2006 the ECtHR concluded that Article 11 had been contravened partly on the basis of an opinion of the European Committee.[119] In the Koua Poirrez case[120] the ECtHR ruled that Article 14 and Article 1 Protocol No. 1 of the ECHR were applicable with explicit reference to a contravention of Article 12 ESC established by the European Committee. In this way, despite the fact that it is not regarded as being 'binding on all persons' by the Dutch courts, the right to social security laid down in Article 12 ESC can still have important legal implications for the domestic courts. The same applies in respect of the conclusions of the European Committee for Social Rights.

6. Direct action against legislation/policy on the basis of the ECHR

On the basis of the ECHR actions are usually brought against concrete, individual (legal) acts *vis-à-vis* a citizen, but for at least the last twenty five years it has also been possible for individuals to complain directly about legislation.[121] This does not mean there is an unlimited right to complain about legislation. The condition stipulated in Article 34 ECHR that the applicant must be a victim, excludes the situation in which action is brought against a law by just anyone or in which the compatibility of a law with the ECHR is judged without a connection to a particular case. The ECtHR summarises the doctrine in the following standard consideration: [122]

> *'The Court recalls that Article 34 (former Article 25) of the Convention does not provide individuals with any actio popularis for the interpretation of the Convention; nor may it form the basis of a claim made in abstracto that a law contravenes the Convention. Nevertheless, the Court has held that Article 34 entitles individuals to contend that a law in itself violates their rights, without any individual measure of implementation, if they are directly affected by it or run a risk of being directly affected by it (...).'*

It even appears that there is an increasing tendency to use the ECHR to test legislation /policy. If a large group is affected or if a large group runs the risk of being affected, the law forms a contravention of the ECHR right being invoked.

[119] ECtHR 11 January 2006, Appl.nos. 52562/99 & 52620/99 (Sørensen and Rasmussen); see also ECtHR 27 February 2007, Appl.no. 11002/05 (Associated Society of Locomotive Engineers & Firemen) and ECtHR 2 July 2002, Appl.no. 30668/96 (Wilson, National Union of Journalists and Others).
[120] ECtHR 30 September 2003, Appl.no. 40892/98 (Koua Poirrez).
[121] See in detail regarding this subject: F.M.C. Vlemminx, Hirst en Broniowski: de tandem waarmee het EHRM de nationale wetgever in de wielen rijdt, NJB 2007, p. 322-329.
[122] ECtHR 22 November 2001 (dec), Appl.no. 45330/99 (S.L. v. Austria).

Below are two examples, derived from the right to property of Article 1 of Protocol No. 1 ECHR.

The first example is the Broniowski case.[123] In simple terms the case is as follows. At the end of the Second World War the border between Poland and the Soviet Union was re-drawn, resulting in the need to repatriate some 1.2 million Poles. Poland accepted the duty to compensate these people for property left behind and laid this duty down in a law. However, 80,000 claimants were left empty-handed and the statutory right to compensation was more or less abolished. Broniowski inherited his right to compensation from his grandmother and invoked Article 1 of Protocol No. 1 of the ECHR. This gave rise to the question of whether Poland had found a 'fair balance' between 'the competing interests of the individual and of the community as a whole', and in particular the question of whether the individual had been saddled with 'a disproportionate and excessive burden'. The ECtHR ruled that:

> *'the Polish State has not been able to adduce satisfactory grounds justifying, in terms of Article 1 of Protocol No 1, the extent to which it has continuously failed over many years to implement an entitlement conferred on the applicant, as on thousands of other Bug River claimants, by Polish legislation.'*

Thus ultimately 80,000 were saddled with 'a disproportionate and excessive burden' as referred to in Article 1 Protocol No. 1. What is remarkable about the Broniowski case, and also the reason why it is being examined in some detail, is that the ECtHR addresses Article 46 ECHR and the problems created in connection with work load when a large group of identical cases, or clone cases, is involved. The ECtHR talks of 'an underlying systemic problem'. It establishes that 167 comparable cases have already been filed against Poland and 'that general measures at national level are undoubtedly called for in execution of the present judgment, measures which must take into account the many people affected.' These measures must include 'a scheme which offers to those affected redress for the Convention violation identified in the instant judgment in relation to the present applicant.'

This Broniowski case was followed in fairly high tempo by other judgements on clone cases and Article 46, each time involving legislation/policy.[124] The Hutten

[123] ECtHR 22 June 2004, Appl.no. 31443/96 (Broniowski).

[124] ECtHR 6 October 2005, Appl.no. 23032/02 (Lukenda); ECtHR 1 March 2006, Appl.no. 56581/00 (Sejdovic 2), following ECtHR 10 November 2004; ECtHR 29 March 2006, Appl.no. 64897/01 (Ernestina Zullo 2), following EHRM 10 November 2004; ECtHR 29 March 2006, Appl.no. 65102/01 (Mostacciuolo 2), following ECtHR 10 November 2004; ECtHR 29 March 2006, Appl.no. 64890/01 (Apicella 2), following ECtHR 10 November 2004; ECtHR 29 March 2006, Appl.no. 36813/97 (Scordino 2), following ECtHR 29 July 2004; ECtHR 19 June 2006, Appl.no. 35014/97 (Hutten-Czapska 2), following ECtHR 22 February 2005.

Czapska case, which once again addressed Article 1 of Protocol No. 1 is worthy of a special mention.[125] This case concerns the statutory rent that landlords may charge tenants. This rent is so low that it fails to cover maintenance costs. A hundred thousand house owners/landlords run the risk of being saddled with a 'disproportionate and excessive burden'. The ECtHR ruled that Article 1 of Protocol No. 1 had been violated and the issuance of instructions based on Article 46 is given a name of its own:[126]

> *'This kind of adjudicative approach by the Court to systemic or structural problems in the national legal order has been described as a "pilot-judgment procedure" (...). One of the relevant factors considered by the Court in devising and applying that procedure has been the growing threat to the Convention system resulting from large numbers of repetitive cases that derive from, among other things, the same structural or systemic problem (...).'*

The workload was reason for the ECtHR to test legislation/policy. Concerns about the workload are understandable. It is estimated that before 2010 some 250,000 complaints will be filed at Strasbourg! The ECtHR is running the risk of collapsing under its own success.

In other words bringing action against legislation and policy on the basis of the ECHR is increasing in importance. This fact is nicely in keeping with the possibility referred to in the previous section of granting social human rights and conclusions of international committees in this field more weight on the basis of the ECHR. After all, the international committees usually consider the compliance with the social human rights in abstracto. Their conclusions are firmly based on the assessment of legislation/policy in the country reports of which the states are proud. What possible failures can be established now on the basis of the ECHR as regards legislation/policy? We will discuss two possibilities here.

In the first place it is conceivable that the law is arbitrary in its effects. This issue is addressed in the Hirst case[127]: in England detainees automatically lose their right

[125] ECtHR 19 June 2006, Appl.no 35014/97 (Hutten-Czapska).

[126] The two examples discussed here derive from Poland, but the proceedings also affected countries such as Italy (see ECtHR 29 March 2006, Appl.no. 36813/97 Scordino 2) and the Netherlands (ECtHR 6 July 2006, Appl.no. 13600/02 Baybaşin).

[127] ECtHR 6 October 2005, Appl.no. 74025/01 (Hirst).

to vote after their sentencing pursuant to a statutory rule.[128] Hirst was convicted of homicide and sentenced to life imprisonment; he filed an objection based on Article 3 of Protocol No. 1 against the loss of his right to vote. Because the law failed to differentiate and thus automatically withdrew the right to vote from all convicted prisoners the ECtHR deemed that the law lacked proportionality. The law is arbitrary in its effects:

> *'The Court recalls that the Chamber found that the measure lacked proportionality, essentially as it was an automatic blanket ban imposed on all convicted prisoners which was arbitrary in its effects (...).'*

The cause of this arbitrariness lies in the lack of differentiation made by the legislator:

> *'The 1983 Act remains a blunt instrument. It strips of their Convention right to vote a significant category of persons and it does so in a way which is indiscriminate.'*

In the second place it is conceivable that the law has an objective but that this objective is not realised. This situation is addressed in the Asmundsson case.[129] Asmundsson had been in receipt of an invalidity benefit since 1978, being unfit to perform his job as a seaman. Because of the Pension Fund's financial difficulties the disability criteria in the law were tightened and the continuation of the right to benefit was made subject not only to disability with regard to the previous job but with regard to work in general. Asmundsson performed onshore work and was thus fit for work in respect of this. As a result, after a transition period of 5 years, his benefit was withdrawn in full. The fact that this amendment to the act affected an extremely small group was deemed by the ECtHR to be irreconcilable with the objective of the amendment, however legitimate this objective may be:

[128] A similar case was heard by the Administrative Jurisdiction Division of the Council of State in 2003, LJN: AM5435. Pursuant to the Dutch Constitution and Elections Act a citizen who is placed under tutelage due to mental disorder also automatically loses his or her right to vote. At the Administrative Jurisdiction Division an appeal was made to Article 25 ICCPR. The Division first conducted an abstract test and concluded, without further explanation, that the exclusion in general cannot be deemed to be an unreasonable restriction. It went on to remark, with respect to the concrete test, that this issue falls outside the jurisdiction of the court. Now, as a result of this ruling by the Division there is a legislative proposal (first reading) (28-2-2006) containing the declaration that there are grounds for considering a proposal to amend the Constitution, with the object of nullifying the provision on the exclusion of incapable persons from the right to vote.; TK 2005-2006, 30 471, nos. 1-3.

[129] ECtHR 12 October 2004, Appl.no. 60669/00 (Asmundsson). In ECtHR 25 October 2005, Appl.no. 584531/00 (Niedzwiecki) a similar proportionality test was conducted. Child benefit was refused because the party involved was only in possession of a special residence permit. The purpose of the act was to restrict the granting of child benefit to those persons deemed to be permanently resident in Germany. Because the permanency of the residence of a person in possession of a special residence permit is unsure, this purpose of the act is not realised. The distinction made here is ruled to be contrary to Article 14 in conjunction with Article 8 ECHR.

> *'The above-mentioned legitimate concerns about the need to resolve the Fund's financial difficulties seem hard to reconcile with the facts (...).'*

Furthermore, in view of the fact that this small group was so hard hit while the great majority continued to receive their benefits as usual implies unequal treatment as referred to in Article 14 ECHR, which weighs heavily when compatibility with Article 1 of Protocol No. 1 is examined. These circumstances, together with the fact that Asmundsson lost 1/3 of his total income while remaining in the same state of health, were reason for the ECtHR to rule that, despite the 5 year transition period, as an individual Asmundsson had been saddled with an 'excessive and disproportionate burden'.

7. Conclusion

This article clearly shows that Dutch constitutional law has the occasional schizophrenic twitch. There is more than enough food available in the Netherlands and yet still the right to adequate food in Article 11 ICESCR is not enforceable in court. This is not only the fate of the right to adequate food. It appears that not a single fundamental social human right can be successfully invoked in Dutch courts. These rights are brushed aside because they are not 'binding on all persons' within the meaning of Articles 93 and 94 of the Dutch Constitution. The rights in the ECHR are all 'binding on all persons' and therefore we should seek to enforce compliance with social human rights on the basis of the rights in the ECHR. This is possible because the ECtHR addresses other relevant treaty provisions when it examines a complaint and does so even if the Netherlands does not regard these other treaty obligations to be 'binding on all persons'. It is hard for the Dutch courts to disregard this approach of the ECtHR. The path via the ECHR is all the more interesting given that the ECtHR is increasingly addressing legislation/policy. On this issue too it is hard for the Dutch courts not to follow the ECtHR. However, expectations should not be too high. Apparently the general opinion in the Netherlands is that the ECHR cannot be contravened when a whole group is affected by the legislation/policy. Violations are only committed in exceptional cases. This is demonstrated by the Salah Sheekh case[130] in which it was ruled that the Netherlands had violated the prohibition of torture and inhumane or degrading treatment in Article 3 ECHR. Salah Sheekh is a Somali asylum seeker. He belongs to a Somali minority group, the Reer Hamar, whose human rights are regularly threatened. Salah Sheekh himself had been beaten and had been the victim of forced labour. His father and brother had been murdered and his sister raped. When, despite all this, the Netherlands still initiated deportation proceedings Salah Sheekh successfully filed a complaint with the ECtHR. The case was a rude awakening for the Netherlands because the ECtHR agreed with the fact that Salah Sheekh bypassed the Administrative Jurisdiction Division of the

[130] ECtHR 11 January 2007, Appl.no. 1948/04 (Salah Sheekh).

Council of State and filed a complaint directly in Strasbourg. Particular attention should be given to the test applied by the Administrative Jurisdiction Division in this type of case against Article 3. The ECtHR examines this test in detail:

> *'the Administrative Jurisdiction Division held in a decision of 7 November 2003 (JV 2004/17) that even if an alien is to be expelled to a country where organised, large-scale human rights violations are committed against a group to which that alien belongs, he or she will have to make out a convincing case that specific facts and circumstances exist relating to him or her personally in order to be eligible for the protection offered by Article 3 of the Convention (...)'.*

The nature of this test was clearly demonstrated in a ruling made by the Division one month later.[131] The Division rejected the appeal to Article 3 although this concerned a Somali woman:

> *'(...) who claimed that she belonged to the Reer Hamar and had been subjected to ill-treatment, including rape. The Administrative Jurisdiction Division considered that the alleged events demonstrated neither that the acts committed had been specifically aimed at the appellant personally, nor that she therefore found herself in a position which substantially deviated from that of other members of the Reer Hamar in Somalia.'*

According to the Division rape is not necessarily personal. How will a Court of Human Rights view this? Is it surprising that the ECtHR does not deem it necessary that Salah Sheekh first seeks relief from the Division? A case must be exceptional in the extreme if an appeal to Article 3 ECHR is to have a chance of success in the Netherlands. If this line of reasoning is let loose on Article 2 ECHR, only the deadest of the dead will successfully be able to invoke Article 2 in the Netherlands! In a nutshell, the traditional resistance in the Netherlands still has to be worn down even under the ECHR. And yet the ECHR remains our best trump simply because of the imperative eye in Strasbourg. The day when our thirst for the recognition of social human rights will be quenched is inescapably approaching.

[131] ABRvS 5 December 2003, JV 2004, 62.

Chapter 4
Veiled justice

The courts' compassionate case law regarding hunger

Bart Wernaart

1. Introduction: charged with rejecting the human right to food

In this book[132] and elsewhere Dutch courts are portrayed as little short of barbaric for rejecting direct applicability of the human right to food. This picture is based on an undeniable rejection in the considerations of the courts. In this contribution I want to revisit the relevant case law asking the question if the rejection of direct applicability of the right to food implies that in the Netherlands hungry people are actually being let down by the courts.

The structure of the chapter is as follows: first I will present the reasons why the courts reject direct applicability of the right to food. Then I will analyse the substantive outcomes of the cases in which the courts pronounce these rejections. This will lead me to the conclusion that the rejection of the right to food mainly concerns the form of the right to food as a human right but not necessarily the need of people to have access to food. Finally, I will examine indications that we are on the brink of a recognition by the Dutch courts of the right to food not only by substance but also in its form as a human right.

2. Reasons for rejecting direct applicability

In rejecting direct applicability of the right to food and other social, economic and cultural rights enshrined in the Dutch constitution and in the International Covenant on Economic, Social and Cultural Rights (ICESCR), Dutch courts duly follow the interpretation given by the Dutch government when it presented the act ratifying the Covenant to parliament.[133] The reasons the government gave for this interpretation were double-edged. On the one hand the government was of the opinion that the nature of social, economic and cultural rights is incompatible with direct applicability. On the other hand they believed direct applicability to be unnecessary because the interests these rights target were sufficiently protected by social security legislation in the Netherlands.

[132] See in particular Chapter 2 by Van der Meulen and Vlemminx and Chapter 3 by Vlemminx.
[133] Parliamentary Documents II 1975-1976 13 932, nr. 3.

This interpretation did not come as a surprise. The Netherlands had already taken a somewhat cautious position during the negotiations preceding the drafting of the treaty that became the ICESCR.[134] Subsequently, the government consistently held on to its interpretation. For example in their third periodic report concerning the implementation of the ICESCR, they state that 'most [ICESCR] provisions cannot be applied directly'[135] and further that 'the Netherlands has a comprehensive system of social benefits guaranteeing its citizens an adequate minimum income.'[136] The few paragraphs dedicated to the right to food convey the impression that a possibility to base a claim directly on the right to food, as laid down in Article 11 ICESCR, is considered unnecessary, due to this system of social benefits and the fact that 'the amount of food available in the Netherlands (...) greatly exceeds domestic demand'.[137]

True to the historic method of interpretation – the method by which the courts give the law the meaning that the legislator intended – Dutch case law adheres to this view. Apart from some summary trial procedures, in which the Courts decided that the question of direct applicability of Article 11 ICESCR is 'too complex' to be answered in such a procedure,[138] or simply ignored the issue,[139] each and every claim invoking Article 11 ICESCR is rejected. To this end the courts employ a phrasing that is repeated time and time again with little or no connection to particularities of the case in point. This phrasing revolves around the following wordings:

> *Article 11 ICESCR is not generally binding as required by Articles 93 and 94 of the Dutch Constitution.*[140] *Furthermore, Article 11 ICESCR by its nature is not directly applicable, for its content is not sufficiently precise to distil concrete rights that can be invoked in a court of law.*

Some variations can be found. For example:

> *From the wording of Article 11 ICESR follows that it does not contain standards the courts can apply directly as yardstick to judge decisions*

---

[134] See UN doc. A/C.3/SR.1266 (1963) para. 57-63. The Dutch government took the position that Article 11 of the ICESCR 'was too detailed, covered many matters which went beyond the competence of the Third Committee, was not consistent with the bald statement relating to the rights to housing and clothing, and was more appropriate to a declaration than to a legally-binding instrument.'

[135] E/1994/104/Add.30, implementation of the International Covenant on Economic, Social and Cultural Rights. Third periodic reports submitted by States parties under Articles 16 and 17 of the Covenant, Addendum, the Netherlands, para. 8.

[136] Idem, para. 329.

[137] Idem, para. 331.

[138] Summary Trail Court Haarlem, 11 July 2007 LJN: BB0998.

[139] Summary Trail Court Haarlem, 29 July 2008 LJN: BE9491.

[140] On the role of Article 93 and 94 of the Dutch Constitution, see the contribution by Van der Meulen and Vlemminx.

> *of administrative authorities, because this provision is insufficiently concrete for such use. Therefore it first needs to be elaborated in national legislation.*[141]

Normally, the Courts deal quickly with a request to directly apply an ICESCR provision, that is, they reject it immediately. This way they hardly provide us with any further explanation beyond the standard considerations just mentioned.

An exception is found in a ruling from 2007, in which the Central Court of Appeal (CRvB) expressed its opinion on the status of some ICESCR provisions. An asylum seeker, awaiting a decision on his application for a residence permit, contested the decision of the local Public Social Welfare Centre to withhold a social security benefit that would enable him to provide for himself and his younger brother. One of the arguments he put forward was that the State has the duty to provide at least some care, because he was residing lawfully in the Netherlands. His Attorney, anticipating the standard considerations with regard to the invoked Articles 9, 11 and 13 ICESCR, argued that in 1986 the Central Court of Appeal had considered with regard to Article 7 ICESCR that it would be incorrect to assume that direct applicability would never be possible.[142] In its response, the Central Court of Appeal seems to consider Article 7 ICESCR on equal treatment of women and man as an exceptional provision in the Covenant. The Court argued in the first place that Article 7 ICESCR[143] is quite concrete, unlike Articles 9, 11 and 13 ICESCR. In the second place, the Court held that the Committee on Economic, Social and Cultural Rights of the United Nations stated in one of their General Comments that some of the provisions of the ICESCR, including Article 7 ICESCR, are suitable for direct applicability. Apparently, the Court is referring to General Comment 3.[144] In Section 5, the Committee on Economic, Social and Cultural Rights lists 6 ICESCR provisions that 'would seem to be capable of immediate application by judicial and other organs in many national legal systems.'[145]

This reference by the Court to General Comment 3 is hardly convincing. Firstly, the list in this General Comment not only mentions Article 7, but also includes Article 13(2)(a)(3) and (4) ICESCR, one of the provisions that were invoked in this case.[146] Secondly, if the General Comments are to carry weight, shouldn't the other General Comments be taken into consideration as well? General Comment 12,

---

[141] This rendering is taken from a ruling by the Council of State of 19 April 2007, LJN BA4289.

[142] CRvB 3 July 1986, TAR 1986, 215. It must be remarked, however, that the CRvB had already withdrawn somewhat from this position in a later ruling in which it held that direct applicability of one of the provisions of the ICESCR would be 'a total exception to the Covenant's general character'.

[143] Article 7 ICESCR contains *inter alia* the principle guaranteeing that women's conditions are not inferior to those enjoyed by men, with equal remuneration for equal work.

[144] Committee on Economic, Social and Cultural Rights, General Comment 3, 1990; UN Doc. E/1991/23. See Annex 2.

[145] The Committee of Economic, Social and Cultural Rights; General Comment 3, para. 5, 1990.

[146] Along with Articles 3, 7(a)(i), 8, 10, and 15(3) ICESCR.

concerning the right to adequate food,[147] aims to inform the Member States of the concrete content of the right to food.[148] Thereby a plea is made for effective juridical or other appropriate remedies.[149] In the light of this authoritative interpretation of the right to food it would require some explanation why the old interpretation by the government is still followed and not considered outdated. Summing up: while the arguments used by the Court are contestable, the message is clear and consistent with the legal practice: there is no direct applicability of Article 11 ICESCR in the Netherlands.

3. The cases

Most scholars who value human rights turn away in disgust from courts who find it in their hearts to write down formulae such as quoted above. However, if we take a closer look at the courts' decisions in their entirety, the situation turns out not to be as bleak as the formula suggests. The case most quoted in this respect, is a case in point.

The ruling is given in a dispute between on the one hand a refugee from Somalia with her 23 month old child and on the other hand the COA.[150] COA is the agency charged with care for and housing of asylum seekers. To this end they operate so-called reception centres. In this case the lady from Somalia contested a decision taken by the COA denying her access to such a reception centre. The reason was that her request for a residence permit had been denied. She was awaiting the decision on a second request. It was COA's policy only to house people awaiting a first decision. In her argument she took recourse to *inter alia* Article 11 ICESCR. She argued that denying her and her child shelter in a reception centre, in combination with the circumstance that she was not allowed to work in the Netherlands, left her without (adequate) means of subsistence and fully dependent on charity help.

The case was decided by the District Court in Amsterdam.[151] The court first rejected direct applicability of Article 11 ICESCR using wordings along the lines quoted above. This did not, however, keep the court from ruling in favour of the claimant. The applicable policy guidelines required COA to take account of

---

[147] The Committee on Economic, Social and Cultural Rights, General Comment 12, 1999; UN Doc. E/C.12/1999/5. See Annex 2.

[148] In General Comment 12, the Committee on Economic, Social and Cultural Rights, responds to Objective 7.4. of the World Food Summit 'To clarify the content of the right to adequate food and the fundamental right of everyone to be free from hunger, and as stated in the International Covenant on Economic, Social and Cultural Rights and other relevant international and regional instruments, and to give particular attention to implementation and full and progressive realization of this right as a means of achieving food security for all.'

[149] The Committee on Economic, Social and Cultural Rights, General Comment 12, 1999; UN Doc. E/C.12/1999/5, para. 32.

[150] Freely translated an abbreviation for: Central Organ for Reception Service for Asylum seekers.

[151] District Court of Amsterdam, 13 March 2001, LJN: AB0942.

distressing humanitarian circumstances. In the light of the circumstances of the case, among them the fact that the claimant did not have sufficient means of subsistence – including food – to provide for herself and her child, the reasoning given by COA did not convince the court that COA had sufficiently taken account of distressing humanitarian circumstances. The District Court of Amsterdam quashed COA's decision, and ruled that COA had to reassess the case.

This approach is in line with the government's reasoning. The lack of direct applicability of the human right to food is made up for by national instruments ensuring protection from hunger.

Not surprisingly, most cases in which Article 11 ICESCR is invoked concern people on low incomes, prisoners, elderly, disabled persons and asylum seekers. Generally, Article 11 ICESCR is invoked to support demands for social benefits, providing the claimant with a minimum means of subsistence. 'Article 11 ICESCR cases' come in two categories. The first consists of the cases in which the level of social benefits is under discussion. In these cases, the claimant enjoys social benefits, but s/he considers these benefits to be inadequate and thus not guaranteeing minimum means of subsistence.[152] Here, the right to food is seldom explicitly an issue: the focus is on Dutch legislation and procedures concerning social benefits. The second category consists of cases in which the claimant has no access to social benefits at all, and, as a consequence, has no means of subsistence that are provided for by the government. The claimants are mostly asylum seekers residing in the Netherlands.[153] A majority of the 'Article 11 ICESCR cases' concern this last category.

Again two situations can be distinguished in which an asylum seeker invokes Article 11 ICESCR. In the first situation, the asylum seeker stays legally in the Netherlands (Article 8 Aliens Act), while awaiting a final decision concerning a residing permit (Article 8f-g Aliens Act 2000), or concerning certain administrative procedures (Article 8h Aliens Act 2000). In this situation, the asylum seekers 'are housed in one of a number of reception centres[154] scattered throughout the

[152] For instance: Central Court of Appeal, 1 November 2005, LJN: AU5600, 9 May 2006, LJN: AX2177, 1 October 2008, LJN: BF4589.

[153] For instance: Central Court of Appeal, 25 May 2004, LJN: AP0561; 8 July 2005, LJN: AT910211; 11 October 2007, LJN: BB5687; 21 November 2007, LJN: BB9625; District Court of The Hague, 30 August 2000, LJN: AA6959; 23 January 2006, LJN: AV0548; District Court of Arnhem, 25 May 2007, LJN: BA6562; District Court of Rotterdam, 19 September 2007, LJN: BB5715; 24 December 2007, LJN: BC0852; District Court Haarlem 8 April 2008 (summary trial), LJN: BD3399; District Court of Amsterdam: 4 August 1999, LJN: AA4043; 13 March 2001, LJN: AB0942.

[154] In this contribution, the term 'reception centre' should be understood as a reception centre for asylum seekers.

country,'[155] but generally have no further rights concerning income support or other social benefits.[156] In the second situation, the asylum seeker stays illegally in the Netherlands, and generally has no access to social benefits,[157] or possibility to stay in a reception centre.[158] The above is based on the so-called 'Koppelingswet', an Act that partly excludes asylum seekers without residence permit from entitlements to general social benefits in the Netherlands. A majority of the cases in which asylum seekers invoke Article 11 ICESCR also deal with the 'Koppelingswet.' The courts generally rule that this Act does not conflict with international law, and is not disproportionate (it is a suitable means to achieve its legitimate purposes).[159] Therefore this Act can be used as a legal basis to deal with the issue at stake. In 'normal' circumstances, this means that an asylum seeker, unlawfully residing in the Netherlands, or lawfully residing but not having a residence permit, cannot make a successful claim to most general social benefits, and thus also cannot make a successful appeal to a right to food.[160]

Remarkably, in some cases the presence of charity-help is held against the claimant: the fact that an asylum seeker received shelter in a care facility owned and operated by a charity organisation, was one of the arguments on which the summary trial court of Haarlem judged that the municipality of Haarlem rightfully rejected an application for housing.[161] Remarkably, in other cases dependence on charity-help, is taken into account by the Courts to establish the severe circumstances the claimant is facing.[162]

The exceptions to the standard legal practice are the most interesting in this context, and may possibly determine whether the Dutch courts let the hungry down in practice.

Above, a case[163] was mentioned in which the court, despite the rejection of direct applicability of Article 11 ICESCR, was willing to consider other arguments to determine whether the claimant should be granted certain benefits. The specific circumstances played a role: the strict policy of the COA with regard to the

[155] Implementation of the International Covenant on Economic, Social and Cultural Rights, Third periodic reports submitted by States parties under Articles 16 and 17 of the Covenant, the Netherlands. E/1994/104/Add.30; 23 August 2005, para. 373.

[156] For instance, District Court of Amsterdam, 13 March 2001 LJN: AB0942.

[157] For instance, Central Court of Appeal, 21 November 2007 LJN: BB9625.

[158] District Court of Amsterdam, 13 March 2001, LJN: AB0942.

[159] Central Court of Appeal, 26 June 2001, LJN: AB2324.

[160] For instance: Central Court of Appeal, 25 May 2004 LJN: AP0561; 8 July 2005, LJN: AT9102; 11 October 2007, LJN: BB5687; 21 November 2007, LJN: BB9625; District Court of The Hague 30 August 2000, LJN: AA6959; 23 January 2006 LJN: AV0548; District Court of Arnhem, 25 May 2007; LJN: BA6562; District Court of Rotterdam, 24 December 2007, LJN: BC0852, District Court of Haarlem 8 April 2008 (summary trial), LJN: BD3399; District Court of Amsterdam: 4 August 1999, LJN: AA4043.

[161] Summary Trail Court of Haarlem, 29 July 2008, LJN: BE9491.

[162] For instance, District Court of Amsterdam, 13 March 2001, LJN: AB0942.

[163] District Court of Amsterdam, 13 March 2001, LJN: AB0942.

interpretation of 'distressing humanitarian circumstances' had to be reviewed in this case, for the situation in which a woman with child does not have entitlements to any means of subsistence, including food, could in all reasonableness not be regarded as not being a distressing humanitarian situation.

Another case revealed that the needs of children present a special circumstance that justifies a kinder approach, compared to the standard legal practice. Remarkably, the Central Court of Appeal took recourse to the International Convention of the Rights of the Child (ICRC). Article 27 ICRC contains *inter alia* the right to food.[164] There had been uncertainty concerning the direct applicability of this provision,[165] for the question had not explicitly been dealt with by the Central Court of Appeal,[166] and the Dutch Government had not mentioned this provision in a list of ICRC provisions they considered directly applicable. The view of the Central Court of Appeal following on implicitly from its ruling on the direct applicability of certain ICRC provisions including the right to food, prompted the Minister of Justice to broaden the scope of persons entitled to certain social benefits: children without a residence permit, but legally remaining in the Netherlands also had entitlements to income support, based on the direct applicability of the ICRC.[167] In late 2007 the District Court of Rotterdam understood these decisions as recognition of direct applicability of the entire ICRC. In this case, after rejecting direct applicability of Article 11 ICESCR, the Court argued that the International Convention of the Rights of the Child is directly applicable, including Article 27. The District Court of Rotterdam held that in a situation in which a child, receiving a monthly allowance from the government, can prove that it does not have sufficient money to provide for himself (*in casu* food and housing), the amount of the monthly allowance should be reviewed.[168]

4. Verdict

The human rights achievements of the Dutch courts in the context of food security may be less bleak then they are portrayed to be. In rejecting direct applicability the courts slavishly follow the lead of politics. This does not mean, however, that the courts deprive the hungry of all rights. If not *by form*, at least there seems to be a right to food *by substance* for asylum seekers who lawfully reside in the Netherlands without a residence permit and find themselves in distressing humanitarian circumstances, or are children. Only the latter is formally recognised.

[164] See Annex 1.
[165] Central Court of Appeal, 24 January 2006, LJN: AV0197. Illegale kinderen en recht op bijstand in het licht van het IVK, mr. C.H. Slingenberg, available at: www.crossborderwelfare.nl/publicatielieneke.doc.
[166] Central Court of Appeal, 24 January 2006, LJN: AV0197.
[167] Wijziging Regeling verstrekking bepaalde categorieën vreemdelingen; Regeling van de Minister van justitie van 22 December 2006, nr. 5458886/06/DVB.
[168] District Court of Rotterdam, 19 September 2007 LJN: BB5715.

For asylum seekers staying illegally in the Netherlands no right to food could be found in the Article 11 ICESCR cases, not by substance and certainly not by form.[169]

So far, the courts in the Netherlands withhold the high status of human right from the entitlements of people to food. This makes such entitlements favours granted by the Dutch legislator, not state obligations the Netherlands is living up to and for which the Netherlands accepts responsibility in the international arena. The courts may lack the courage to depart from the opinions of politics, but they do not seem to lack compassion for people in need. There seems to be some justice behind the veil of the Dutch rejection of human rights.

The example of the recognition of the right to food in the International Convention of the Rights of the Child as a directly applicable right, shows that in situations where the courts do muster the courage to serve justice over politics this may have a profound impact on politics.

5. A new day dawning?

The Central Court of Appeal in its ruling on the ICRC may indicate that the time is ripe for a recognition of the right to food, not only by substance but also as a human right. Recent developments show that if Dutch politics makes the move, some courts at least may be willing to follow. On the 3rd of March 2008 the Dutch Minister of Foreign Affairs Maxime Verhagen made a speech to the Human Rights Council:[170] Verhagen proudly announced *'here today that the Netherlands will join the group of countries who have recognised the right to water as a human right.'*

Shortly after this proud announcement and explicitly referring to it, the local Court of Heerlen recognised direct applicability of Articles 11 and 12 of the ICESCR in a case where a water provider (with practically a monopoly in the province) shut down the water supplies to a customer, whose payments were overdue. The Court ruled that in doing so the water provider violated the customer's right to water and health as codified in Articles 11 and 12 of the ICESCR. Article 11 ICESCR was thus directly applied. This was done on the initiative of the court itself (no Article 11 ICESCR claim was made by the customer).[171]

It remains to be seen if this decision will hold in appeal and if the justice establishment will muster the same human rights-awareness. The Central Court

[169] As there is in Switzerland: BGE 121 I 367, 27 October 1995, available at: www.servat.unibe.ch/law/dfr/c1121367.html#Opinion.
[170] See: www2.ohchr.org/english/bodies/hrcouncil/7session/hls/Netherlands-E.pdf see also: www.minbuza.nl/en/news/speeches_and_articles,2008/03/Minister-Verhagen-s-speech-at-the-Human-Rights-Cou.html and www.minbuza.nl/en/news/newsflashes,2008/03/Verhagen-at-the-UN-Human-Rights-Council.html.
[171] District Court of Maastricht, (Section Heerlen), 25 June 2008, LJN: BD5759.

of Appeal has not yet risen to the challenge, but it may have overlooked the new position of Minister Verhagen. On the first of October 2008, the Central Court of Appeal again rejected the direct applicability of Article 11 ICESCR by once again using the same standard consideration.[172]

If indeed we are witnessing a turn in the tide resulting in the formal recognition of the right to food as an enforceable human right by the courts in the Netherlands, that may even have consequences for a group that has so far been completely left out: aliens not legally residing in the Netherlands. Under the ambit of a directly applicable right to food, the responsible authorities will at the very least have to substantiate why withholding all subsistence from these people constitutes a justifiable limitation of their human rights.

[172] CRvB, 1 October 2008, LJN: BF4589.

Chapter 5

Hunger as a policy instrument?

Some reflections on workfare and forced labour

Gijsbert Vonk

1. Introduction

The title for this treatise, which I was invited to reflect upon, is a provocative one. The notion of 'hunger as a policy instrument' implies that states consciously use the deprivation of people to realise certain public goals. Perhaps such policies are conceivable in periods of (temporary) [173] loss of civilisation but otherwise? Modern civilised states adhere to the rule of law. They have accepted socio-economic fundamental rights, such as the right to a decent standard of living and the right to social security. These rights presuppose that nobody deserves to be poor and that states are under a duty to create a welfare system, in whatever form, which is aimed at alleviating the needs of the masses, without threatening the livelihood of others. Should we really question the sincerity of the efforts of states in this field? God have mercy on the cynics!

But suppose we would approach the phenomenon of hunger not so much from the point of view of official policies, but rather from the function it has in crafting relations in our society, as some sociologists want us to do? Then it is true that it can be an easy instrument in the hands of some to force others (the poor) into certain behaviour. Even though official policy would not readily formulate it in these terms, the fact remains that the threat of taking away the livelihood of social security beneficiaries can be used as a disciplinary tool. Indeed, in literature this disciplinary function of social welfare law, in particular the application of the so-called 'work test' in some social security schemes (in particular unemployment benefits and social assistance) is widely accepted.[174]

In this contribution I touch upon the instrumental perspective of poverty in our social welfare systems (withholding benefit rights to enforce certain behaviour). Firstly, I will describe how the previously existing harsh and unforgiving attitude towards poverty has given way to a rights-based approach. This is reflected in the state of the law which attempts to strike a careful balance between the right to benefit and the work obligations of the beneficiary. Such a balance can, for

[173] Conceivable? I will never forget my disbelief when I read for the first time that somewhere in the machinery of Nazi Germany a plan had been deployed to literally starve the slave people in order to create *Lebensraum* for themselves.
[174] Cranston, R. *Legal foundations of the welfare state,* Weidenfeld and Nicholson, London, 1995, 202-207.

example, be illustrated with reference to the concept of 'suitable employment' used in various social security acts. (Section 2) Secondly, I will demonstrate how the balance between rights and duties runs the risk of being disturbed as a result of the modern *workfare* approach, adopted by an increasing number of states. This approach combines a strict work test and sanctions with far-reaching discretionary powers for the administration. If handled in the wrong way, this may result in repressive practices which ignore the right to social security for the individual claimant (Section 3). Thirdly and finally, I will argue that governments, and in particular courts have a duty to maintain the balance between rights and obligations, by giving effective application to the fundamental right to work and the prohibition of forced labour (Section 4).

In literature workfare policies are very often discussed within the context of the welfare state at large. The concept of the welfare state is wider than the concept of social security. In the first place 'welfare' refers not only to the provision of income security in case of poverty or certain risks, such as unemployment and old age, but to the whole spectrum of government action intended to make sure that citizens meet their basic needs, such as education, housing and health. In the second place 'welfare' refers not only to cash benefit schemes, but also to various types of services and in-kind programmes often considered to fall outside the social security domain, such as probation and parole, child protection services, socialisation services, etc.[175] As I am mostly focussing on the work test as a condition for receiving unemployment and social assistance cash benefits, this chapter is confined to the subject of social security only.

Although the questions and issues raised will be dealt with in a general manner, the situation in the Netherlands is taken as a fixed point of reference.

2. Social security and work as fundamental rights

In the pre-modern period there was no need for society to worry too much about activating the poor to work. Hunger drove the masses into activity. Only with the emergence of the first public relief schemes did work incentives have to be organised. Thus, the nineteenth-century poor laws made a clear cut distinction between the deserving and the undeserving poor. Those who were not incapacitated as a result of sickness, handicap or old age (the so-called able-bodied) were forced to participate in publicly organized employment. Work houses were set up in which men, women and children had to carry out manual activities in miserable conditions for long hours a day. There was no easy escape from the work house. Dealing with poverty was considered to be part of the policing function of the state. Vagrancy was a criminal offence. Sometimes vagabonds were literally rounded

[175] For the meaning of the term social welfare, see P.R. Popple and L. Leighninger, Social work, social welfare and the American Society, Boston, Allyn & Bacon (1993).

up and kept in confinement in forced labour camps.[176] Especially during the 19th century, a period when the state had largely withdrawn from society and many traditional forms of care had eroded under the influence of the industrial revolution, poverty and poor law dependency were a terrible ordeal for the people involved.[177]

During the course of the 20th century conditions improved. Work houses were abolished and the notion of public charity gradually eroded in favour of legal guarantees for the beneficiary. After the Second World War these developments culminated in the recognition to social security and work as human rights. Nowadays we believe that nobody deserves to be poor, that everybody should be able to earn his living in an occupation freely entered upon (the right to work) and that there should be a system for the protection of major risks ensuing from labour and life and poverty in general (the right to social security).

The *right to work* is adopted in the national constitutions of many countries, as well as in various international instruments on socio-economic fundamental rights, such as the European Social Charter (ESC[178]) and the International Covenant on Social, Economic and Cultural Rights (ICESCR[179]). It is not easy to catch the meaning of this right in a single phrase. On the one hand it presupposes a positive obligation of the state to strive for a high and stable level of employment and to provide and promote employment services and occupational training. On the other hand it displays the characteristics of a freedom right where it protects the freedom of occupation. In the latter sense the right to work is related to the prohibition of slavery and forced labour, adopted in the other human rights instruments, such as the International Covenant on Economic, Social and Cultural Rights (Art. 6) the European Convention on Human Rights (Art. 4) and conventions of the ILO[180].

The *right to social security* is not one-dimensional either. It is generally understood that it presupposes a final responsibility for the state to set or orchestrate a system of income protection for a number of social risks, such as unemployment, sickness, invalidity, old age, etc. and a basic floor for persons who are left without any other

[176] A Dutch example of these practices was the poor colony of Veenhuizen in the province of Drenthe, which started in the first half of the 19th century as a charitable initiative, but quickly evolved into a forced labour camp in which vagrants were locked up.
[177] For an historical account *cf.* Samuel Fleischacker, *A short history of distributive justice*, Harvard University Press, Cambridge (Mass.) 2004.
[178] Art. 1 ESC.
[179] Art. 6 ICESCR.
[180] E.g. Forced Labour Convention, 1930 (ILO, nr. 29).

resources.[181] Also the rule of law is considered fundamental for the right to social security as a human right. The previously existing system of charity and poor laws did not presuppose a legal obligation to provide benefit. This was a matter of discretion for the institutions which are – at most – under a moral obligation to deliver. Consequently, in a charitable system there cannot be any corresponding right to a benefit either for the recipient. The right to social security, however, presupposes a system under which persons are entitled to support. This suggests that the beneficiary has some sort of legally defined position.[182] The inference of a legally defined position is that claims should be vested in the law and can be enforced.

When we combine the right to work and the right to social security, it follows that it is no longer acceptable to require a person to offer his physical capacities to some form of organised employment in direct return for public support. Indeed, when we look at our present day social security systems it appears that the relation between benefit entitlement and duty to work is no longer construed in this way. A person qualifies for benefit when he or she satisfies certain objective conditions and then has to accept certain obligations. These obligations may very well involve a duty to work and accept offers of employment, but as a starting point, the beneficiary is allowed to choose the employment himself on the free labour market, thereby taking into account the job offers that are made to him by the labour exchange or social security authorities. Moreover the concept of employment is a qualified one. It refers to *suitable employment*. It has been argued that the right to work should be interpreted as a right to obtain suitable employment and that work undertaken under pressure of need (the withdrawal of benefit) is not suitable within the meaning of the relevant standards of the ILO.[183] In my eyes this might be too optimistic, but at least the term suitable employment presupposes that the work must somehow be fitted to a person's capacities, experience, education, etc. Whether or not this is the case should be established on the basis of individual merits. Finally, the sanctions which may be imposed if a person refuses to accept suitable employment are to be regulated by the law. They should be proportionate

[181] Cf. Eibe Riedel 'The human right to social security, some challenges' in: *Social Security as a Human Right, Drafting a General Comment on Article 9 ICESCR*, Eibe Riedel (ed.), Berlin 2007, 16-28. On the nature of the state obligations in the field of social security. cf. Gijsbert J. Vonk & Albertjan Tollenaar, 'The public interest and the welfare state, a legal approach, Working paper annual legal research network conference, Groningen, 2008: www.rug.nl/crbs/onderzoek/networkconference2008/papers2008.

[182] This fundamental distinction between charity and social security has been reflected upon in an impressive opinion of Advocate General *Mayras* of the European Court of Justice in the Frilli-case, where the court had to decide on the exclusion of 'social assistance' from the material scope of social security Regulation no. 3/58 (now Regulation 1408/71) and is – as a bottom line – still accepted in the Court's case law to this day. ECJ 22 June 1972 (Frilli), 1972 *ECR* 457.

[183] Cf. Frans Joseph Smit 'Suitable employment as a human right', in: *Reflections on International Law from the Low Countries,* Erik Denters and Nico Schrijvers (eds.), Kluwer Law International, 1998, 191-197 (as a summary of the author's valuable doctoral thesis, *Passende arbeid als recht van de mens, recht op arbeid en op werkloosheidsuitkering in internationaal en Nederlands recht,* Kluwer 1994).

and the claimant should have access to an independent judge if he or she wishes to dispute the decision of the administration.

If we accept this description as an abstract model for applying the work test in modern social security systems, it follows that a harsh work test is replaced by a more moderate system which tries to strike a balance between the right to benefit and obligation to work

3. Workfare and the effect on the balance between rights and duties

The balance between rights and duties in our social welfare system is, however, shifting. More emphasis is being placed on the participatory function of social security and the enforcement of rules while social security's function in protecting income is pushed to the background. These changes, which have particularly affected the area of social assistance, are often referred to as the 'workfare' approach. [184] In a more narrow sense of the word the term workfare is reserved for the situation in which claimants of social assistance are required to perform work, often in public service jobs, as a condition of receiving aid. The latter, narrow definition immediately brings us to the heart of the matter: does workfare reintroduce the 19th century practice of requiring 'a person to offer his physical capacities to some form of organised employment in direct return for public support', as I phrased it above?

The answer to this question is not necessarily affirmative. As will be pointed out below, much depends on the legal guarantees surrounding the workfare programmes. But before we embark on this subject, let us first of all have a look at how the workfare approach might upset the balance between rights and obligations that has gradually developed in our social security systems.

3.1 Indiscriminate powers over the individual

One of the characteristics of the legislative changes which are introduced under the influence of the workfare approach is the erosion of the legally defined position for the beneficiary in favour of more discretion for the local authorities responsible for benefit payments. Thus, for example, with respect to the new Dutch *Work and Assistance Act* (2003) municipalities have been given wide-ranging powers in the area of the activation of beneficiaries and the imposition of sanctions. The powers are partly regulated in the form of local council regulations and policy guidelines, but sometimes powers are directly placed in the hands of individual administrators. The purpose of the *Work and Assistance Act* is to provide tailor-

[184] For an overview cf. Joel F. Handler, *Social Citizenship and Workfare in the United States and Western Europe, The Paradox of Inclusion*, Cambridge studies in law and society, Cambridge, 2004.

made solutions. The obligations or sanctions to be imposed upon the individual claimant are to be established on a case by case basis; each claimant should get the treatment that he or she deserves. In this way social assistance has once again become a matter of unilateral, discretionary judgment of the authorities involved. In the meantime work first projects have sprung up all over the place. In these projects young people, new claimants and sometime also those who have been on benefit for a long period are obliged to carry out services for the community, in the parks, in work factories or as street cleaners. Those who do not join in will be confronted with benefits cuts.

The danger of wide-ranging discretionary powers is simply that these powers may be applied too rigorously. Local councils, or indeed individual administrators, may hold extreme views on the duties of the claimants of benefit and impose unfair demands and disproportionate sanctions. If the powers of the councils or administrators are not properly counterbalanced by the legal rights of the beneficiaries, this might give rise to forms of abuse. For example, what should be made of the suggestion put forward in 2005 by the former Dutch Minister of Social Affairs, De Geus, that prostitution should be accepted as suitable employment? Or the Enschede local authority that threatened to force women on social assistance to join a dating agency in the hope that they would catch a man with money of his own? These examples refer to incidents which were intensively debated but never became reality. But one has to remain vigilant. Just Google the Dutch terms for 'social assistance' (bijstand) 'forced labour' (dwangarbeid) and one is bombarded with hits which tell stories of claimants of social assistance who feel degraded, bullied or exposed by the workfare practices. However subjective these stories may be, they do testify to the fact that the balance between rights and duties has shifted.

3.2 Exclusionary effects

The second danger presents itself on the level of society as a whole. To what extent do workfare policies really contribute to the emancipation of the poor? This question has been emphatically raised by the French born sociologist *Loïc Wacquant* in a book called *Punishing the poor, from welfare state to penal state.*[185] In this book the author gives a disturbing insight into some developments in the US.

The American system of social assistance was reformed in 1996 by an act with the ominous title *Personal responsibility and work opportunity reconciliation act*. The initiative for the reform came from the conservative politician Clay Shaw but the act was signed by Bill Clinton. The reform created a synthesis between the Democratic

[185] The book as advertised under this title by the author but – to the best of my knowledge – a version in the English language has not yet appeared. There is French version and a translation into Dutch under the title *Straf de Armen, het nieuwe beleid van de social onzekerheid*, Berchem, 2006.

belief in activating policies and the Republican dogma of personal responsibility. The changes in the American system are similar to those that were introduced in the Netherlands, but they were earlier, more radical and on a grander scale. The system operates on the basis of block grants paid out by the central government to the various states. The grants are only payable when the states satisfy certain conditions: benefit dependency may not last longer than five years, a certain percentage of the beneficiaries must have found paid employment, and lack of co-operation on the part of the claimants must be punished with benefit cuts.

Wacquant is outspokenly negative about the new American social assistance system. According to the author one half of the claimants finds employment in temporary low paid jobs which condemn the persons involved to a permanent state of poverty. The other half simply disappears out of the system and will be adopted in the anonymity of the urban slums. According to Wacquant, this is symptomatic of the changes which are occurring in the US. Problems are no longer solved on the basis of a social agenda. Instead the citizen is made fully responsible for his own life and the degree to which he or she can participate in society. Where these policies fail the state reacts with sanctions and criminal measures. Since 1970, the number of prisoners has more than quadrupled (442%). This rise mostly involves small-scale offenders, drug addicts and the homeless. The majority originate from the poorer segments of society. Most of them are black. One out of three black males would be subject to some form of criminal surveillance. It is as if Dickensian times are returning: the prison house as an alternative for social security. In the meanwhile maintaining the repressive system requires a huge public investment. Thus, according to the author, the 'light' American liberal state has developed a 'heavy' substructure to suppress the poor.

In the Netherlands, evaluations of the new social assistance legislation that came into force in 2003 show the same sort of ambivalence as in the US. On the one hand they are positive because more people have found paid employment. On the other hand there are concerns about the quality of the jobs involved and there is a similar problem concerning the beneficiaries that simply no longer appear on the records; what happened to them?[186] Another parallel relates to the repressive nature of the workfare policies. The public tolerance for those who do not adhere strictly to the rules and regulations has plummeted and the intensity of the sanctions (in number and severity) has increased. The dark side of the policies are mostly ignored. For example, hardly any attention is paid to research outcomes which point out that most of the fraud is committed unwillingly, due to a lack of understanding of the exact nature of the rules and bureaucratic obligations.[187]

---

[186] Cf. Els Sol, Julie Castonguay, Hanneke van Lindert en Yvonne van Amstel, *Work first werkt. Op weg naar een evidence based-work first*, Utrecht, 2007.

[187] Cf. Frank Miedema and Bob Post, *Fraude Onderzocht. Oorzaken en achtergronden van fraude met AOW, Anw en AKW*, ITS, Nijmegen, 2001.

In the meantime also in the Netherlands the number of prisoners has exploded; quadrupled since 1984. And again we are not dealing with heavy criminals but with representatives of the marginal groups in the society, such as the homeless, problematic youths and foreigners.[188]

In my view, these developments should be critically monitored, because, however biased Wacquant's analysis may be, it does contain a warning. The warning is that, if pursued too ruthlessly activation policies designed to be beneficial for the citizens may have the opposite effect, condemning 'those who stay behind' to a permanent underclass. Incidentally, this is also the conclusion of a comparative study on workfare policies in the US and Europe by Joel F. Handler which repeatedly points at the risk that workfare policies have an exclusionary effect for those who are least employable.[189]

4. Workfare or forced labour?

Despite the dangers described above, it is important to bear in mind that the introduction of workfare policies does not necessarily imply a re-introduction of 19th century repressive practices. In the first place according to the general workfare philosophy forced public employment is not a goal in itself, but a first step on the ladder towards participation in the normal labour market. One could argue that the more successful workfare programmes are in realising an effective flow to regular employment, the more justified it is to make claimants perform tasks as a first step towards activity. In the second place, workfare practices do not operate in a void, but within the context of the law. This infers that claimants have a legal remedy when they feel that their rights are being infringed. It is then up to the courts to restore the balance between the rights and obligations. As long as the rule of law is upheld the dangers of the workfare approach described in the previous section can be kept at bay. This presupposes an active attitude of the courts and a strict focus upon the protection of human rights. After all, it is the judge who must eventually test the legality of benefit cuts and sanctions and it is case law that must provide a framework of criteria for judging the validity of obligations imposed on the beneficiaries.

In order for the courts to exercise their corrective powers workfare practices should come within the realm of the law in the first place. That they do, seems perfectly obvious, but in reality this is not always recognised. Thus, for example, I came across an article in the New York Times, dealing with a decision of the U.S. Equal Employment Opportunity Commission of 30th September 1999 in which

[188] M. Boone and M. Moerings, 'De Cellenexplosie, Voorlopig gehechten, veroordeelden, vreemdelingen, jeugdigen en tbs', *Justitiële verkenningen*, 2007, 9-30.
[189] Joel F. Handler, *Social Citizenship and Workfare in the United States and Western Europe, The Paradox of Inclusion*, Cambridge studies in law and society, Cambridge, 2004, 245.

it was ruled that the New York administration had violated federal law when it turned away women who said they were being sexually harassed while working for their public assistance benefits.[190] The allegations were very serious but the city had maintained that the women were not employees and had no legal right to protection from sex discrimination in the workplace, so it refused to co-operate with the proceedings. The case perfectly illustrates the harshness of the new policies and the responsibility of the courts (or as the case may be: independent tribunals) to restore the balance between rights and obligations.

The right to work (in particular the freedom of occupation) and the prohibition of slavery and forced labour as contained in several international human rights instruments may play an important role in providing a framework for testing the legality of workfare practices. But curiously there are hardly any national or international cases in which concrete decisions of social security administrations to withhold benefit rights were considered to be in violation of any of these rights, at least not that I am aware of. The general understanding seems to be that work duties may be imposed as a benefit condition and that withholding benefit rights does not impede someone's freedom of occupation, let alone constitute forced labour. Thus, in the case of Johan Henk Talmus v. the Netherlands the (then) European Commission on Human Rights ruled that benefit cuts that were imposed because the applicant had refused to look for suitable employment, did not infringe Art. 4 ECHR (prohibition of slavery and forced labour). 'In order to qualify for unemployment benefits (...) the applicant was required to look for and accept employment which was deemed suitable for him. Since he refused to comply with this condition, his benefits were temporarily reduced. It does not appear, however, that the applicant was in any way forced to perform any kind of labour or that his refusal to look for employment other than that of independent scientist and social critic made him liable to any measures other than the reduction of his unemployment benefit. In these circumstances, the Commission cannot find that the present complaint raises any issues under Articles 4 para. 2, 9 and 1 of the Convention'.[191]

Similarly, in 2002 a Danish court did not accept that a suspension of benefit upon a refusal to accept an activation offer, was contrary to Art. 4 ECHR. The Danish welfare-to-work option (activation offer) that the person had received was considered to be fair, in the sense that he was able to carry out the work, and that the objective of the requirement for participation in the welfare-to-work programme as a condition for receiving cash benefits was to become self-supporting. The only consequence of the plaintiff's refusal to participate in the welfare-to-work programme was that he, while the job offer was open, lost his entitlement to cash

[190] *New York Times*, front page, Nina Bernstein, 1 October 1999.
[191] ECtHR no. 30300/96, decisions of 26 February 1997, J.H. Talmon v. Netherlands, EHRLR 1997, 448-449.

benefits. According to the court there was neither evidence in the wording of the ECHR, nor in the case law of the European Court of Human Rights to assume that the connection between participation in a welfare-to-work programme and the reception of cash benefits could imply a threat of 'punishment' within the meaning of Art. 4 of the Convention.[192]

Likewise in a recent judgement the local Amsterdam court ruled that the duty of a social assistance claimant to work in a sweet factory did not constitute forced labour, because the applicant was not coerced into anything, even when threatened with having his benefits suspended. Moreover, the work duty had to be considered as a 'normal civic obligation' as being one of the exceptions referred to in Art. 4.[193]

I have great trouble in accepting the way courts tend to reject outright the relevance of the prohibition to forced labour in social security cases. Firstly, by doing so courts fail to appreciate the great responsibility which rests upon them to protect the proper balance between rights and obligations in times of the introduction of workfare policies. Secondly, the case law does not recognise that withholding benefit rights may constitute a serious form of pressure and coercion upon the person involved. According to the European Court of Human Rights forced labour is labour exacted under menace of any penalty and performed against the will of the person involved, that is work for which he has not offered himself voluntarily.[194] I fail to see why under some circumstances, particularly long-term benefit dependency, sanctions would not amount to such a penalty.

And finally, the case law still seems to be based on the underlying assumption that nobody is forced to apply for social security benefits. But in my view the point of reference for judging the question of whether freedom rights are violated should not be the freedom to apply for a benefit, but rather the right to social security as a fundamental right for all.

5. A refreshing approach: the district court of Arnhem, 8 October 2008

I would welcome a more critical approach to be adopted in case law. Referring to the general criteria which have been developed by the European Court of Human Rights in respect of the probation of forced labour and compulsory employment contained in Art. 4 ECHR, in particular in the cases of *Van der Mussele* and *Siliadin*[195] it should be possible to extract a number of minimum standards for

[192] High Court of Eastern Denmark, B-1420, 19 December 2002 discussed in Danny Pieters and Bernhard Zaglmayer, *Social security cases in Europe: national courts*, Antwerp, 2006, 45-46.
[193] Rechtbank Amsterdam 8 June 2008, LJN: BD7415, Rechtbank Amsterdam, AWB 07/3717 WWB, AWB 07/4337 WWB, AWB 07/4609 WWB.
[194] ECtHR judgement of 23 November 1983, Van der Mussele v. Belgium, para 34.
[195] ECtHR judgement of 26 July 2005, Siliadin v. France.

testing the legality of workfare practices. There are three elements which in my view should be tested critically, i.e.:

- the nature of the employment;
- the duration of the employment;
- severity of the sanction.

With regard to the *nature of the employment* a distinction must be made between labour which is carried out under a regular contract of service and labour in public employment in consideration for the payment of benefits. When dealing with the second type of labour, special attention should be paid to the question of whether the labour suits a person's physical and mental capacities and is in line with the official policy objective of workfare, i.e. it should be conducive to one's personal development and the improvement of one's chances on the regular labour market. Degrading or useless activities shall not be allowed.

With regard to the second element of the *duration*, I am of the opinion that a person should not be required to carry out public duties on a more or less permanent basis. When, after a couple of years the person in question has still not found a regular job, the civic work duties should be ended to allow a person to choose voluntary activities or sheltered employment.

Finally, with regard to *sanctions*, it may be argued that a refusal to accept work duties may only be penalised by temporary or partial benefit cuts. Full and permanent benefit cuts shall not be permitted.

Just before submitting this article the Dutch district court of Arnhem passed a judgement which must be considered a major breakthrough in this matter.[196] The case dealt with a social assistance beneficiary with an academic background who had been told to accept certain activities, offered to him by a 'training centre', a facility set up under the work first programme of the town of Arnhem. The beneficiary was told to sign a 'job experience agreement' under which he was given the choice either to work as a public gardener (weeding, hoeing), or to pack boxes of super glue. The beneficiary had signed the agreement but subsequently refused to co-operate in the activities imposed on him by his 'case manager'. This resulted in a penalty of a 40% benefit cut, during the period of one month. The beneficiary appealed against this penalty in an attempt to force the court to take a principle stance about the work first duties in relation to the prohibition of forced labour. In its judgement the court came to the conclusion that the practices of the local council of Arnhem were not contrary to the prohibition of slavery and forced labour contained in Art. 4 ECHR. The fact that the workfare activities were not voluntary because imposed under the threat of a penalty, did not alter this conclusion because, according to the court, social assistance is merely a safety

[196] Rechtbank Arnhem, 8 October 2008, LJN BF 7284.

net which presupposes that a person will return to paid employment as soon as possible. But while on the one hand the court ruled that in this case the activities offered should not be considered as disproportionate and excessive, it did on the other hand envisage that work first practices may run contrary to Art. 4 ECHR, i.e. in the case of a beneficiary who is forced to carry out activities under threat of a penalty for a longer time when it is clear that such activities are in no way conducive to re-integration in the regular labour market. Eventually the court ruled that the penalty imposed by the Arnhem council was unlawful, because the council had failed to make clear that the activities in question could have a positive impact upon the job opportunities of the person involved, leaving aside whether there was a breach of Art. 4 ECHR or not.

As said, for our subject this judgment of the Arnhem court constitutes a major breakthrough. Not only is it a novelty that a court admits that forcing beneficiaries into activities in return for benefit payments may constitute forced labour or compulsory employment, but we also begin to see the first contours of the criteria to which benefit agencies engaged in work first practices must adhere.

6. Conclusion

The harsh work test as applied in the 19th century has gradually softened. As a consequence of the right to work and the right to social security, it has been replaced by a more moderate system which tries to strike a balance between the right to benefit and the obligation to work. This balance is, however, shifting as a result of the workfare approach. This approach, generally advocated to activate social security claimants, is not devoid of dangers. Social security administrations may exercise indiscriminate powers over individuals. And, if pursued too ruthlessly, activation policies which are designed to be beneficial for citizens may have the opposite effect, condemning 'those who stay behind' to a permanent underclass. It is up to the courts to restore the balance between rights and obligations. As long as the rule of law is upheld the dangers of the workfare approach can be kept at bay. This requires an active attitude of the courts and a strict focus upon the protection of human rights. When dealing with the prohibition of forced labour both the European Court of Human Rights and some national courts tend to apply these rights too restrictively. I have argued that it is no longer acceptable to reject outright the relevance of the prohibition of forced labour in social security cases. Instead, minimum standards for testing the legality of workfare practices should be developed. These should refer to the nature and duration of public employment duties, as well as to the severity of the sanctions. The recent decision of the Dutch district court of Arnhem of 8 October 2008 constitutes a perfect illustration of such an alternative approach.

Chapter 6
The Netherlands and the Making of the Voluntary Guidelines on the Right to Food

Arne Oshaug

1. Introduction

It is not possible to report on the behaviour of a specific country during the negotiations of a very important human rights document such as the Voluntary Guidelines on the Right to Food (hereafter Right to Food Guidelines), without including some analysis of the eighteen month long intergovernmental process[197] in 2003 and 2004, and some reflections on the hindsight view of some of the events leading up to the formation of an international organisation like the UN. The situation before the establishment of the UN was unthinkable atrocities against humanity. The misery in the 1930's in the USA and Europe and the events and consequences, including the abuse of power, of the Second World War were devastating. Those events, one may say, led to the creation of the United Nations.

Two people in particular turned out to be very important for the events in the period until human rights were born. These were President F.D. Roosevelt, and his wife Eleanor Roosevelt, both with visions and political wisdom of rare proportions. In his inaugural speech in 1937 President Roosevelt said:

> *'The test of our progress is not whether we add more to the abundance of those who have much; it is whether we provide enough for those who have too little.'*

In his inaugural speech in 1941 he presented the famous four freedoms, freedom from want, freedom from fear, freedom of speech and freedom of faith. He was persuaded to take another turn as the president of the USA because of the special situation created by the World War II, and in his inaugural speech in 1944 he stated that:

> *'We have come to a clear realization of the fact that true individual freedom cannot exist without economic security and independence. 'Necessitous men are not free men.' People who are out of a job are the stuff of which*

[197] IGWG: Intergovernmental Working Group for the Elaboration of a Set of Voluntary Guidelines to Support the Progressive Realization of the Right to Adequate Food in the Context of National Food Security.

> *dictatorships are made. In our day these economic truths have become accepted as self-evident.'*

In 1948 the United Nations launched a far-reaching project for the recognition and realisation of human rights by adopting the Universal Declaration of Human Rights (UDHR). It encompasses economic, social and cultural rights. These rights were not universally recognised in 1948, and far from universally respected – they were rights to be achieved through national and international efforts as reflected in the UDHR (Eide, 2005). The awareness and recognition of civil and political rights improved greatly through the 1990's in particular, albeit in the face of severe challenges due to ethnic conflicts. After the attack on the World Trade Center in New York this awareness increased even more. To some extent this has led to a war on terror, with some known cases of serious violations of basic human rights (Eide, 2005). Economic, social and cultural rights as described in the UDHR and further specified in the International Covenant on Economic, Social and Cultural Rights have been less accepted as 'real human rights', but some academics started to work on giving content to the right to adequate food.

An important international event which provided incitement to further development was the World Food Summit (WFS) in 1996, with FAO providing the secretarial function. There the members of the WFS agreed to clarify the content of the right to adequate food (Rome Declaration and Plan of action, Commitment 7, Objective 4(e)).

However, less than a year after the Summit, a Draft International Code of Conduct on the Right to Food as a Human Right became available, drafted by a small group of non-governmental organisations (NGOs) and endorsed, by September 1997, by more than eight hundred NGOs (Oshaug, 2005).

It soon became clear that a proposal for a binding instrument would have no chance of being adopted. But already at the preparatory negotiations before the World Food Summit in 1996 it became evident that even the term 'code of conduct' would be strongly rejected by several delegations. Therefore a different strategy was adopted: namely to request 'a better definition of the content of the human right to adequate food' (Eide and Kracht, 2005). With such a call coming out of the Summit there would be time in which to make further efforts.

The interim period saw the birth of a number of initiatives and important documents inspired by the Draft Code of Conduct (Oshaug, 2005; Oshaug and Eide, 2003). The World Food Summit: *five years later* (WFS:*fyl*), held in Rome 12-16 June 2002, provided another opportunity to address the issue of a code of conduct on the right to adequate food. The negotiations during the preparatory week 3-9 June in 2002 on a consensus text for the Summit were intense up to the last minute, with one issue still not agreed upon at the formal closure of that meeting: i.e.

whether to include a recommendation for a voluntary code of conduct for guiding the implementation policies and programmes aimed at achieving the human right to adequate food, or just one focusing on food security (as defined by the WFS in 1996). The United States did not want any reference to the human right to adequate food. This country was also against starting a process towards a code of conduct, and never changed in substance their suggestion for the text of this paragraph during the negotiations.[198] USA was partly supported by the European Union (chaired by Spain). To cut a long story short, the decision was made to establish an Intergovernmental Working Group, to elaborate, in a period of two years, a set of voluntary guidelines to support Member States' efforts to achieve the progressive realisation of the right to adequate food in the context of national food security (for further details see Oshaug, 2005).[199]

At the formal meeting of the WFS:*fyl* on Monday 10 June, the High Commissioner for Human Rights, Mary Robinson, in her statement at the Summit proper presented the main features of the General Comment 12 (Robinson, 2002).

Many stakeholders considered the agreement resulting in paragraph 10 in the document coming out of WFS:*fyl*, which included establishing an intergovernmental working group on elaborating a set of voluntary guidelines, a lost cause. They thought that agreeing on 'voluntary guidelines' was tantamount to giving up the fight for a useful tool on how to make human rights based policies to fight hunger, food insecurity and malnutrition. Others thought that, on the contrary, this was in fact a victory since the process of continuing the work on the right to adequate food was now assured, and General Comment No. 12 on the right to adequate food would provide a very good basis for that continuation.[200]

This was the situation when the Intergovernmental Working Group on the Right to Food (IGWG) met for the first time on 25 March 2003 in Rome. Within less than two years the working group would present agreed guidelines to FAO's Committee on World Food Security (CFS) before their final approval by the FAO Council in

[198] USA's proposal 8 June 2002: 'We believe that voluntary guidelines may help countries to develop and strengthen policies directed at achieving food security. We ask the CFS to consider an appropriate mechanism to elaborate such guidelines, bearing in mind the importance of full stakeholder participation including civil society and the private sector and the need for consensus decision-making on such guidelines.'

[199] Paragraph 10 of the final Declaration states: We invite the FAO Council to establish at its One Hundred and Twenty-third session an Intergovernmental Working Group, with the participation of stakeholders, in the context of the WFS follow-up, to elaborate, in a period of two years, a set of voluntary guidelines to support Member States' efforts to achieve the progressive realisation of the right to adequate food in the context of national food security; we ask the FAO, in close collaboration with relevant treaty bodies, agencies and programmes of the UN system, to assist the Intergovernmental Working Group, which shall report on its work to the Committee on World Food Security.

[200] In fact the only objective of the WFS fulfilled at the time was Objective 7.4, which materialised as General Comment No. 12 to the ICESCR.

the autumn of 2004. The special feature of this decision was that, for the first time in history, a tool intended to help countries to formulate human rights based policies was to be negotiated in an intergovernmental setting.

Prior to the first meeting, the secretariat, through a coordinating *ad hoc* unit established by FAO for the IGWG process, had prepared a synthesis report of submissions from a large number of stakeholders (FAO, 2003a). This report points out that the 123rd Session of the FAO Council (28 October–01 November 2002) decided to establish the IGWG. This gave a mandate to the Secretariat of the Working Group to prepare a synthesis report of the submissions sent by governments and stakeholders for the first session of the IGWG. This report was presented to the first session of the IGWG as a basis for the deliberations. It was clear then that the contribution from the member countries was modest. The function of the synthesis report was to show the various inputs to the IGWG. The idea was that this would facilitate the discussions. Primarily the purpose of a future Right to Food Guideline was that countries would get access to a practical tool or road map to assist in the implementation of existing legal obligations regarding the right to adequate food and in pursuit of the goals established by the WFS, the Millennium Summit and other major international conferences. The document should thus serve to guide administrative and legislative agendas and systematically identify legal and policy measures and programmes to achieve the realisation of the right to adequate food in all countries (see Oshaug, 2005, for further elaboration).

Many submissions underlined the need to ensure transparency, accountability, decentralisation and participation in food security policies and interventions, and to assist in coordination between different levels and institutions of governments. Furthermore, the Guidelines should be clear and simple, worded in such a way that all stakeholders could identify with them, and they should add value to existing tools. Most submissions stressed the importance of GC12, and some also noted the relevance of the recent General Comment No. 15 on the right to water[201]. There was a general agreement that the Guidelines should be comprehensive, and cover all categories of obligations, i.e. to respect, protect and fulfil the right to adequate food. The need to address long-term development, self-reliance and the immediate needs of those presently hungry or malnourished, was also widely acknowledged (FAO, 2003a).

This, then, was the basis on which the IGWG would start its long and difficult work on elaborating the Guidelines that were to be finalised in September the following year. As the work progressed it turned out that those heavily engaged in the negotiations took on a considerable responsibility, in keeping the process focused and on track so it would finally be a useful practical tool for governments in the fight against hunger, malnutrition and poverty. In this context the role of

[201] CESCR 2002.

the Netherlands became very important, in particular in the period when it held the presidency of the European Union.

2. The intergovernmental process in the IGWG between 2003 and 2004

2.1 The first IGWG meeting

The task of the IGWG was to follow up decisions of WFS:fyl on the Right to Adequate Food (RtF). The philosophy of the negotiations was that '*Nothing is decided until all is agreed*'. In principle this meant that any of the country delegations could stop or delay the process (for more details of the process see Oshaug, 2005). This is against the FAO Constitution where there is an opening for voting, but it is in agreement with most negotiations within the UN system that usually one goes for consensus. The consequence is that most documents negotiated in such a setting would be watered down, because various countries could manipulate the content and not give up until they were satisfied. The concern in such situations is not for the dire situation of the poor and hungry, but for the consequences for domestic policy of each and every member country.

The Synthesis Report (FAO, 2003a) from the secretariat[202] established by FAO provided a good basis for the discussions of the first meeting of the IGWG. The opening statements by the various delegates were a good indication as to who supported useful, practically oriented guidelines, and those who thought that such guidelines were not needed or wanted, and thus would try to prevent their development. There was cautious optimism among supporters that Mr. Tony Hall was recently appointed as the US Ambassador in Rome, because he had earlier advocated for a convention on the human right to food. He opened the negotiations of the IGWG by taking the floor on behalf of the USA as the first country, by explaining what the USA was doing to fight food insecurity both internationally and domestically; he also made a reference to the Universal Declaration of Human Rights and that the right to food was mentioned there, and that the USA recognised the right to have access to food. He underlined that the right to food should therefore be interpreted as a goal and not a human right. Thus a rights-based approach should not be indicated in a possible future guideline. With the USA in the Bureau responsible for drafting a text based on consensus, this was not a good signal to the IGWG member countries, but rather an ominous omen for the possibility of getting guidelines at all.

[202] FAO had established an *ad hoc* secretariat for supporting the Bureau and the IGWG. It was composed of very able staff: Mr. Julian Thomas and Coordinator, Mr. G. Pucci, with the support of FAO Legal Council, Ms. M. Vidar, FAO Legal Office and Mr. F. Mischler, FAO Economic and Social Department. Carlos Lopez, Human Rights Officer, was seconded to the Ad Hoc secretariat from Office of the United Nations High Commissioner for Human Rights (OHCHR).

After the first day the secretariat on behalf of the Bureau drafted a non-paper on *convergence* and *divergence* categories. This added to the issues already recorded in the Synthesis Report (FAO, 2003a); the NGOs and UN agencies also came with important constructive suggestions and comments. Many of the regions made good progress in coordinating their viewpoints. It turned out, however, that the European Regional Group did not function well.

Most of the time set aside for regional consultations was spent by the European Union on their internal coordination. Greece had the EU Presidency at that time and thus functioned as a Chair for the EU countries. The consequence was that at best only a few minutes were left to inform the rest of the European countries about the EU position. The representative of the European Region in the Bureau did his best to consolidate the various viewpoints, but frequently ended up presenting and defending the EU position. Throughout the work the IGWG Chair (the Iranian Ambassador) seemed a bit confused and referred to the EU when he in fact meant to refer to the European Region as a whole. The Chair of the European Region did thus not manage to talk fully on behalf of the Region. The Eastern European country candidates for EU membership in particular did not want to do anything to annoy the Chair of the EU and the EU Commission. It was therefore natural that some of the countries outside the EU, such as Switzerland and Norway, started to collaborate and act together, and presented several common interventions (Oshaug, 2005).

Controversial issues soon became apparent, including the international dimension, justiciability, whether the guidelines should be for states only or also for other actors, and food as a human right in crisis and conflicts. Some countries insisted that there should be no reference to the General Comments, not even to the GC12, no use of human rights language and no indication of human rights principles. These issues turned out to be controversial throughout the negotiations.

The Secretariat in their analysis presented in the Synthesis Report showed immediate optimism about achieving a consensus on several human rights issues related to food, an optimism that was refuted. There was no consensus on very many issues among the most vocal contestants, and at best only the viewpoints of those countries and stakeholders that were interested enough to send contributions to the Secretariat established for the negotiations were reflected. The prospects for achieving the mandate given to the IGWG by the WFS:*fyl* via the FAO Council, seemed bleak from the beginning.

2.2 The second IGWG meeting

In the second meeting of the IGWG, the Bureau presented a first draft of a possible VG (FAO 2003b). However, to the surprise of the participants no negotiations of the text took place. Only presentations of viewpoints and suggestions for changes were allowed. Many felt that at least the introduction and some difficult topics could have been negotiated initially, so as to be sure of what the really difficult issues were. The Bureau[203] maintained that given the newness and complexity of the draft as it was, the wide-ranging opinions about them, their political sensitivity and the limited time available for the negotiation, it was better to have a structured, systematic, and disciplined approach (FAO, 2003c).

Furthermore, the Bureau maintained in the same minutes that the purpose of this second meeting of the IGWG was to identify and agree to broad areas of convergence and divergence, to address opposing views where opinions differed, and to establish mechanisms for negotiating the draft text outside the IGWG meetings. The discussions in the Bureau were apparently very difficult. The representatives that were best prepared and directed much of the negotiation were those representing North America. All the negotiations were conducted in English hampering the active participation of those not fluent in that language (particularly those from a Francophone culture). Many delegations questioned the wisdom of this approach and said that the IGWG would lose a lot of time by not starting with the negotiations, in particular on issues where consultations with the government sector in the home country would be needed.

It was acknowledged that the time given for developing the VG in just three meetings of the IGWG, was probably too short. It was therefore suggested that whenever regional FAO meetings were held, IGWG/VG would be included in the agenda. It is unclear to this author whether this has happened systematically, but it seems to have had a limited impact on the process.

The Bureau suggested an interesting new feature in view of UN negotiations. An open-ended in-session working group was established, chaired by the Chairman of the IGWG. It would include only four spokespersons from each of the seven geographical FAO regions, three representatives from NGOs/CVOs and representatives of UN agencies and intergovernmental organisations. The regions were to appoint spokespersons. The reason given for this arrangement was to facilitate an efficient exchange of views (FAO, 2003d). In a UN setting it is unusual to let other countries speak on behalf of one's own country unless that has been formally agreed otherwise.[204]

[203] Between the first and second meeting of the IGWG the Bureau met seven times to deal with procedural and other matters related to the VG. However, in order to elaborate the first draft of the VG the Bureau met only twice (once in July and once in September) (FAO, 2003c).

[204] Examples of such arrangements where one spokesperson can speak on behalf of several countries are G77, GRULAC (Group of Latin American countries), Middle East Region, African Region, and Nordic Countries.

While most of the regions seemed to manage this arrangement well,[205] it created considerable problems for the European Region. Initially the European Union demanded all the seats for the European Region because they represented 25 countries (including 10 new applicant countries). They maintained that they wanted Italy (Chair) and Ireland (upcoming chair), and representatives of the Commission of the EU as spokespersons. Switzerland and Norway pointed out that the European Region consisted of 44 countries, and both Switzerland and Norway were not EU members but contributed financially to the IGWG. They had considerable interest in the process and were not ready to let other countries speak on their behalf. Gradually also Bulgaria, Croatia, Romania, San Marino and Turkey supported Switzerland and Norway in their demands. Within the EU there were also countries (in particular Germany[206]) that voiced support for letting non-EU members taking part in the informal discussions in the open-ended in-session working group. The result after long negotiations was that the European Region would be represented by Italy and Ireland, with Switzerland, Norway and Germany alternating on the remaining two seats.

The preparatory negotiation within the groups was intense. A considerable problem for the Chairman was controlling the discussion and making it more efficient. The responsibility for being effective and controlling the process was transferred to the regional groups, and in fact excluded many countries from taking an active part in the plenary negotiations. It also made it possible for the more interested and active countries to be more influential and having their viewpoints better reflected in the final text. In fact given the culture of consensus, such an approach restricted participation even further, and gave the sceptics and those negative to the human right to adequate food an advantage. One country could simply block the negotiations by saying that a suggested text was unacceptable.

This was the case in the European Union, where Germany was a proactive supporter of the human right to adequate food, while countries like UK, Denmark and Sweden, many times supported by the EU Commission, were sceptical and sometimes directly blocked consensus within the EU group on certain issues. The Netherlands were not very visible or active outside the EU-group in this process. In such a situation the text had to be re-negotiated to find compromises or simply to be deleted. The many bracketed paragraphs in later summary reports confirmed this.

Features of blockage surfaced. The controversial issues mentioned above remained (Oshaug, 2005). The resistance to having references to the General Comments was equally strong, particularly from the USA and Canada. The developing countries

[205] Maybe except for Asia which included a wide range of developing as well as developed countries. There were disagreements many times between the poorer Asian countries and Japan.
[206] Germany had considerable interest in the process, and had contributed financially and thus made it possible for FAO to manage the process of developing the VG.

wanted to have a reference to official development assistance (ODA), and increased access to global markets for food and agricultural commodities, while the richer countries pointed out that there are forums for these kinds of negotiations and they should not be dealt with in the IGWG.

It was argued that the VG should not establish any new obligations, and some pointed out that the VG should not be weaker in substance than relevant existing documents. This was agreed, but several delegates, NGOs/CVOs and representatives from UN agencies[207] maintained that the draft text from the Bureau was actually a step backwards compared to existing international human rights norms. Some underlined that it must be recognised that the VG cannot make voluntary what is already obligatory.[208] Many maintained that the levels of obligation (respect, protect, fulfil) had to be included,[209] while the USA underlined that they were not prepared to accept any human rights language in the VG. Many delegations argued that a rights-based approach should be adopted,[210] and that the VG should reaffirm the universality, indivisibility and inter-relatedness of all human rights, and be consistent with and reflect relevant provisions of Human Rights Law.[211] It became apparent that the situation was becoming increasingly complex with every subsequent meeting.

2.3 The IGWG inter-sessional meeting

It was clear already during the second meeting of the IGWG that at least one extra meeting was necessary for getting a negotiated agreed text within the time-frame given by the FAO Council in its 123rd Session in 2002. The Bureau decided therefore to have an inter-sessional meeting of the IGWG in February 2004.

The compilation of text proposals from earlier meetings gave a relatively good picture of the diversity of the discussion (FAO, 2004a). The arguments were numerous as to the structure and content of the foreseen document (Oshaug, 2005), including the human rights principle and reference to basic human rights instruments, specification of clear purpose/objectives, a clarification of the relationship between the right to adequate food and food security (essentially GC12), categories of state obligation,[212] and a part dealing with international dimensions.

[207] Jamaica/CARICOM (Caribbean Community) States, Venezuela, Switzerland, Norway, OHCHR, FIAN.
[208] Norway, Senegal, OHCHR, FIAN.
[209] Chile/GRULAC, Mexico, Norway, Special Rapporteur on the Right to Food.
[210] Afghanistan, Norway, Philippines, Special Rapporteur on the right to Food, CESCR, FIAN.
[211] Italy/EU, and acceding countries, Norway, Senegal, CESCR, OHCHR, FIAN.
[212] The USA did not recognise any value of, nor acknowledge, voluntary guidelines, and stated that references to human rights principles, or categories of obligation, could not be accepted because that was human rights language. Again they maintained that the guidelines should be about food security.

In view of the very large number of proposals received, and the fact that no negotiations had taken place up till this inter-sessional meeting, the Bureau was delegated with the authority to consolidate the proposals without changing them. It was stressed that the Bureau[213] would have no negotiating or decision-making authority and that it should seek to ensure that proposals made would be reflected in the document (FAO, 2004a, for further discussion see Oshaug, 2005).

2.4 The third IGWG meeting

The third formal session[214] took place from 5th to 10th July 2004, although it was supposed to finish on the 9th July. However, given the amount of work, many doubted that the time devoted during this week would suffice, even with sessions every evening.

The purpose of this 3rd IGWG meeting was to negotiate an agreed text of the VG. The Bureau had produced a second draft of the VG even though there had been no negotiations up to this meeting (FAO, 2004b). Many considered therefore that the basic documents for the negotiations were the first draft and the subsequent reports from the earlier meetings, including the Compilation Document (FAO, 2004a), and written comments sent to the secretariat by member governments. The secretariat had also provided background papers on issues relevant to the topic.

The first controversy was organisation of the work. Three working groups were established: Working Group I, chaired by the first Vice-Chair of the IGWG, Ambassador Christian Monnoyer (Europe), was to deal with the Preface, the Introduction and Guideline 15 on Natural and Human-made Emergencies. Working Group II, chaired by Mr. Noel De Luna (Asia), was to deal with Part V, the International Framework. Working Group III chaired by Ambassador Mohammad Saeid Noori-Naeeni (Chair of the IGWG), was to negotiate the remaining Voluntary Guidelines (Parts II to IV, excluding VG 15). A 'Friends of the Chair Group'[215] was to be convened as required (FAO, 2004c).

[213] The Bureau had changed in composition. Mr. Moussa Bocar Ly had taken the place for Africa, and the representative for the Pacific (New Zealand) had withdrawn from the Bureau and did not attend IGWG meetings.
[214] The attendance had gradually increased and was about the same as at the inter-sessional IGWG meeting. Delegates from 95 Members of FAO and the United Nations, observers from the Holy See and the Sovereign Order of Malta, one United Nations Agency, two representatives of the United Nations Special Rapporteur on the Right to Adequate Food, two representatives of the CESCR, two intergovernmental organizations, and 12 international non-governmental organizations. Regional meetings and a Bureau meeting took place on 4th July. Again there was a change in the Bureau when Mr. Robert Harris was replaced by Mr. Richard W. Behrend (USA).
[215] This is a frequently used format in UN negotiations. It is an informal group, which can be given a mandate to negotiate certain difficult issues, but is expected to report back in plenary to the rest of the delegates before final formal decisions. It is often composed of the chairperson and main opponents.

While the Working Groups were open-ended, regions were encouraged to limit themselves to a maximum of six country spokespersons from each region, with the possibility of rotating these spokespersons. In addition, delegations were encouraged to conduct negotiations on the basis of regional positions. The advancing of new proposals was discouraged. The support of at least two regions was required to consider new proposals. Proposals were regarded as 'new' if they had not featured in the second drafted VG from the Bureau (FAO, 2004d) or in the Compilation Report (FAO, 2004a).

Now the bitter fruit of delayed negotiation and lost time was felt. There was an attempt to make up for lost time with strict organisation and chairing,[216] by limiting the number of speakers, and by urging everybody to show goodwill, a positive attitude, and restrain interventions to few and short statements.

The negotiations on the agreed text went well, with a good and positive spirit in all the working groups. The attitude of the USA delegation changed completely, from having uncompromising standpoints and arguments, to being more supportive, suggesting compromises where they could accept different solutions, and being more open and listening to arguments from other delegations.[217] On the negative side there was a considerable change in the attitude and willingness to compromise by the EU. The chair of the EU (the Netherlands) displayed a rigid position on several issues, in particular on Part V – the International Dimension. The Chair of the Working Group II pointed in particular to the negative attitude, rigidity and lack of political will of the EU. This created considerable problems for the negotiations in this group.

The negotiations broke down at about 02:30 hours on Saturday morning. The major reason was the rigidity of the Presidency of the EU (the Netherlands[218]) on the International Dimension, and the unwillingness of the G77 to accept compromise formulations by USA[219] and EU. To quote the Chair of the IGWG, nothing was approved before everything was approved. He also stated in the

[216] In addition to the working groups mentioned a number of small fast-working groups were formed to address specific paragraphs in order to find an alternative text. That approach worked well on many occasions.

[217] The people in the US delegation were different. Apparently the hardliners of earlier delegations were replaced by professionals in negotiation, with a strategy, which appeared similar to a proposed approach by Fisher *et al.* (1991).

[218] The Netherlands was coached by UK, Sweden and the EU Commission, all known to be skeptical about the human right to adequate food. Germany on the other hand, which had been one of the steady and firm supporters (both financially and orally), did not engage in providing advice to the Netherlands during the negotiations.

[219] The USA suggested deleting many formulations in Guideline 15, such as references to 'human made', any reference to 'occupation', 'water', mention of international humanitarian law linked to responsibility of occupying powers, 'ongoing conflicts' and abiding by rules governing humanitarian assistance and protection of humanitarian personnel (FAO, 2004d).

Plenary on Saturday noon that with a little more political will, an agreed text could have been achieved. The negotiations had basically collapsed.

3. The role of the Netherlands

This situation of a collapse of the negotiations cannot be detached from the role of the European Union. From the perspective of rich countries, and for the outcome in general, one could say that the EU had many positive suggestions for improving the content of the draft guidelines as they progressed through the various stages of negotiations. A particular challenge to the EU Presidency was needed to coordinate the viewpoints of the various EU member countries so that they had a clear and agreed statement by the Presidency to the rest of the IGWG. When the EU group had finished their coordinating meetings, usually in the morning before the formal sessions started, there was no proper time for discussions with non-EU members. This created a considerable problem and frustrations in delegations of non-EU member countries such as Switzerland and Norway. At one point the EU suggested deleting the Nutrition Guideline from the draft VG since apparently those who suggested it believed in the trickle-down effect. After strong protests from some country delegations and NGOs that suggestion was withdrawn.

The role of Presidency of the EU (and thus the chair of EU-member countries during the negotiations in the IGWG) was demanding. The complex issues presented above show that a clear strategy and capacity were needed. That included an in-depth knowledge of what the right to adequate food means and implies, how that is related to the European Convention on Human Rights, a clarification about what could be accepted as content in a potential VG, what should be the negotiating positions, what should be the non-negotiating positions, who to link up with 'friends' or like minded groups/countries, alternative suggestions for text (shared with other member countries or groups), positions on being a broker, insights into other positions on human rights elsewhere, national government/State obligations[220], and being an internal broker who could find collective compromise texts where member EU countries disagreed. It was very important to have negotiable flexible positions and not just rigid non-negotiable positions.

Often during the negotiations the EU played a broker position to find solutions to difficult questions which threatened to derail the negotiations. The problem with that role was that certain EU countries were hostile or negative to the whole idea of having guidelines on the human right to adequate food. It seemed that they did not look for ways as to how the work of the IGWG could be supportive of the European Convention on Human Rights. The issues seemed to be more about looking for what could not be accepted, and being as negative as possible to the

[220] The fact is that it is the national governments/States who ratify international human rights treaties, and not the EU as a body.

idea of the human right to adequate food. The negative countries seemed to have easy access to the Presidency, were listened to and had a direct impact on the positions presented by the EU Presidency. On the other side of the spectrum of negative, non-interested or indifferent countries was Germany who was a strong promoter of the VG. Germany was in the EU group and must thus have felt both frustrated and sidelined in many situations.[221]

The role of the Netherlands changed during the various phases of the total period of the negotiations. In the beginning the country took part in IGWG as an EU member. As mentioned above, during that period the Netherlands had a low profile in the EU group and did not distinguish themselves by taking any form of internal leading role. This author observed that there were limited contacts between EU member countries, Norway or other countries outside the EU. When the Netherlands took over the Presidency the role of the Netherlands changed. The country became the Chair of the EU delegation together with the EU Commission. That was a demanding role, in a very complex situation. It meant dealing with a complex issue requiring knowledge as to what the human right to adequate food actually meant, using that knowledge to link the negotiations to the European Convention on Human Rights, using that knowledge to guide the EU-member countries in internal discussion, and presenting that in plenary. Furthermore, there is an implication that this knowledge and insight can be used to negotiate, and not only as a statement that was agreed in the EU group during the preparatory meetings, which is what actually happened. The risk of failure was therefore considerable.

3.1 The international dimension

As described above the negotiations broke down during the third meeting of the IGWG. The major reason identified here and stated by others during the negotiations was the rigidity of the Presidency of the EU (the Netherlands[222]) on the International Dimension, and the disagreement between G77, USA and EU. Many wondered how this could happen, with supportive countries like Germany, Norway and Switzerland in the European region.

Many would maintain that there was no real negotiation and coordination within the European group but only within the EU Group, so it was considered that Europe lost out to those countries that were against having an agreed document (VG). It seems that a lack of real communication and coordination led to fragmentation that almost destroyed the entire work of the IGWG. The lack of an agreed text

[221] This is an assumption by the author.

[222] The Netherlands was coached by UK, Sweden and the EU Commission, all known to be skeptical about the human right to adequate food. Germany on the other had, which had been one of the steady and firm supporters (both financially and orally), did not appear, as observed by this author, to be engaged in providing advice to the Netherlands during the negotiations.

on the international dimension,[223] which developing countries had such strong viewpoints on, had brought the negotiations to a standstill. The leadership of the Netherlands holding the Presidency had proved to be inadequate. The complexity of the situation was simply too difficult for the EU negotiator, and the capacity of the EU Presidency to negotiate seemed inadequate in such a complex situation.

3.2 The solution – the 'Friends of the Chair' Meeting, September 2004

Because the negotiations could not be completed during the third and last formal planned meeting of IGWG, it was decided to continue the negotiations parallel to the meeting of the CFS that would take place in Rome from 20th to 23rd September 2004. A last possible formal meeting was planned in October that same year, since the issues left over from the third IGWG were considered so difficult that another final meeting would be necessary before the Guidelines could be presented to the FAO Council for final approval in November 2004. These issues included the international dimension, whether armed conflicts should be included in the same guidelines that addressed man-made emergencies and complex emergency situations, and the rule of law[224].

The time between the third IGWG and this meeting had been well used. The negotiators seemed well prepared, had clarified their positions, and as it turned out were willing to compromise. At the start of the negotiations at the Friends of the Chair Meeting, the EU[225] presented a possible solution to the international dimension. Their suggestion was to simplify the structure of the Voluntary Guidelines into three parts: Section 1 – Preface and Introduction, Section 2 – Enabling Environment, Assistance and Accountability, that would include a last guideline to function as a chapeau to Section 3, entitled International Measures, Actions and Commitments. This suggestion was accepted by the G77 group, creating a very positive atmosphere for the rest of the negotiations. All parties involved accepted a modified text by the G77 for the new and final guideline (Guideline 19). The draft VG was accepted by the CFS (FAO, 2004f).

This last session was blessed by the fact that all the participants had reflected well before coming to the meeting. Much time was spent on mutual backslapping, charm was employed by most of the active discussants, and the initial strongest opponent, the USA, continued to act as a broker and at the same time made sure that certain phrases such as 'appropriate and in accordance with domestic law'

---

[223] A guideline on the international dimension was supported by several industrialised countries and in particular the Civil Society Organisations.

[224] These issues were negotiated in smaller groups in which Syria, Canada, Cuba, USA, Switzerland, Brazil and the Red Cross participated. The proposed guideline on rule of law (16bis in the last draft – FAO, 2004d) was at the suggestion of the African Group moved to Guideline 1, which got the title Democracy, Good Governance, Human Rights, and the Rule of Law (FAO, 2004e).

[225] Their main negotiator this time was Hans van Omen from the Netherlands.

were included in the text at the last minute. No one dared oppose this in the final minutes because of the relief everyone felt about finally getting an agreed text. The charm and last words from the USA remains in stark contrast to the first statement from Ambassador Hall and the fierce resistance to have any guidelines regarding the human right to adequate food at all.

3.3 What can be learned from this experience?

International negotiations are demanding, particularly these, which were the first of their kind in which a human rights instruments was developed in a bilateral setting. The complex issues that the negotiations dealt with demanded a clear strategy and capacity[226]. It was clear from the total process that these negotiations did not fit the model of changing the main negotiator every 6 months as was the case for the EU[227].

It seems also that an in-depth knowledge of what the human right to adequate food means and implies, and how this is related to the European Convention on Human Rights, would be a very hard act to fulfil for a changing presidency. In such a function it would be very hard to demand an adequate capacity of a rotating actor. There should be specific demands as regards the position of EU Presidency for an in-depth knowledge of the content of relevant international human rights instruments and how that could or should be reflected in a potential VG. As it turned out that was either not understood or actively opposed. As a minimum one should realise that one cannot make voluntary what are already recognised obligations for the State party to the ICESCR.

The question as to whom to link up with appeared to mean only delegations of EU members or member-to-be countries. One got the impression that there were contacts with the North American delegations, but in Europe such contacts outside the EU were limited and not actively pursued. Alternative suggestions for texts which could be shared with other member countries or other active groups was not shared for the sake of finding a compromise text, but as a courtesy to show others the position of the EU. To change such a position would require new negotiations within the EU group. The EU today consists of both rich and poor countries. Internally in the Union there is a policy of solidarity, but in this setting the solidarity dimension was international and thus beyond Europe, and that seemed to be unacceptable.

[226] Capacity in this setting should be defined as used in Sabatini (2005): Motivation and acceptance of duty, authority, resources (economic resources, human resources, and organisational resources), capability to communicate, capability for rational decision-making and learning from experience.
[227] The Presidency, i.e. the Presidency of the Council of the European Union, is held by each Member State in turn for a period of six months. During this time, the Presidency is the 'face and voice' of the European Union, speaking on behalf of all Member States. (www.eu2007.de/en/The_Council_Presidency/What_is_the_Presidency/index.html).

There does not seem to be any form of flexibility, which would have been necessary for the Presidency to take on a role as negotiator in this situation. Thus the ability to be a broker, with insight into other delegations' positions with the purpose of finding a compromise text, was simply a theoretical possibility. Signals from the EU Presidency and texts intended to be part of the VG soon materialised as strict non-negotiable positions without any room for manoeuvre.

It could be said that the EU is not suitable for chairing such negotiations and the EU member countries and the EU Commission should have realised this before the negotiations started. The obligations and responsibilities for protecting and promoting human rights rest with the States' governments. They are the parties to the Covenant and not the EU.

In such situations it is simply not enough to have a Presidency that functions as 'the face and voice' of the European Union. Each member that has ratified the ICESCR must be allowed to speak on its own behalf.

References

CESCR (1999) General Comment 12 – The right to adequate food (Art. 11). The Committee on Economic, Social and Cultural Rights, twentieth session. Geneva.

Eide, A. (2005). The importance of economic and social rights in the age of economic globalisation. Chapter 1 in: Barth Eide, W. and Kracht, U. eds. Food and human rights in development. Volume I: Legal and institutional dimensions and selected topics. Intersentia, Antwerp and Oxford.

Eide, W.B. and Kracht, U. (2005). The right to adequate food in human rights instruments: Legal norms and interpretations. Chapter 4 in: Barth Eide, W. and Kracht, U. eds. Food and human rights in development. Volume I: Legal and institutional dimensions and selected topics. Intersentia, Antwerp and Oxford.

FAO (1996). World Food Summit (WFS. www.fao.org/wfs/homepage.htm).

FAO (2003a). Synthesis Report. Intergovernmental Working Group for the Elaboration of a Set of Voluntary Guidelines to Support the Progressive Realization of the Right to Adequate Food in the Context of National Food Security. First Session, Rome, 24-26 March. Food and Agriculture Organization of the United Nations, Rome (IGWG RTFG 1/2).

FAO (2003b) Voluntary Guidelines to Support the Progressive Realization of the Right to Adequate Food in the Context of National Food Security. Draft prepared by the IGWG Bureau for consideration at the Second Session of the IGWG. Rome, 27-29 October 2003. Food and Agriculture Organization of the United Nations, Rome (IGWG/RTFG 2/2).

FAO (2003c) Minutes. Right to Food Guidelines, Bureau meeting, 17 October. Mimeo. Food and Agriculture Organization of the United Nations, Rome.

FAO (2003d) Report of the Chair. Intergovernmental Working Group for the Elaboration of a Set of Voluntary Guidelines to Support the Progressive Realization of the Right to Adequate Food in the Context of National Food Security. Second Session, Rome, 27-29 October. Food and Agriculture Organization of the United Nations, Rome (IGWG/RTFG2/REP).

FAO (2004a) Compilation of text proposals for the Voluntary Guidelines to Support the Progressive Realization of the Right to Adequate Food in the Context of National Food Security. Inter-sessional Meeting of the Open-ended Working Group. Rome, 2 – 5 February. Food and Agriculture Organization of the United Nations, Rome (IGWG/RTFG3/REP2).

FAO (2004b) Second Draft of Voluntary Guidelines to Support the Progressive Realization of the Right to Adequate Food in the Context of National Food Security. Draft proposal by the IGWG Bureau for negotiating at the Third Session of the IGWG. Appendix to the Report of the Chair on Bureau meetings, 26-27 April 2004. Food and Agriculture Organization of the United Nations, Rome (IGWG/RTFG 4/2).

FAO (2004c) Report of the Chair. Intergovernmental Working Group for the Elaboration of a Set of Voluntary Guidelines to Support the Progressive Realization of the Right to Adequate Food in the Context of National Food Security. Third Session. Rome 5-10 July. Food and Agriculture Organization of the United Nations, Rome (IGWG/RTFG 4/REP 1).

FAO, (2004d). Voluntary Guidelines to Support the Progressive Realization of the Right to Adequate Food in the Context of National Food Security. Text approved during IGWG III (5-9 July 2004). Annex to the Report of the Chair of the third session. Food and Agriculture Organization of the United Nations, Rome (IGWG/RTFG 4/REP 1).

FAO (2004e) Report of the Chair. Intergovernmental Working Group for the Elaboration of a Set of Voluntary Guidelines to Support the Progressive Realization of the Right to Adequate Food in the Context of National Food Security. Fourth Session. Rome, 20 – 23 September. Food and Agriculture Organization of the United Nations, Rome (IGWG/RTFG 5).

FAO (2004f) Report. Intergovernmental Working Group for the Elaboration of a Set of Voluntary Guidelines to Support the Progressive Realization of the Right to Adequate Food in the Context of National Food Security. Fourth Session. Rome,23 September. Food and Agriculture Organization of the United Nations, Rome (IGWG/RTFG 5/Rep 1).

Fisher, R., Ury, W. and Patton, B. (1991) Getting to yes. Negotiating Agreement Without Giving In. Penguin Books, New York.

Oshaug, A. (2005) Developing voluntary guidelines for implementing the right to adequate food: anatomy of an intergovernmental process. Chapter 11 in: Barth Eide, W. and Kracht, U. eds. Food and human rights in development. Volume I: Legal and institutional dimensions and selected topics. Intersentia, Antwerp and Oxford.

Oshaug, A. and Eide, W.B. (2003). The long process of giving content to an economic, social and cultural right: twenty-five years with the case of the right to adequate food. In: Bergsmo, M. (ed.) Human rights and criminal justice for the downtrodden. Marinus Nijhoff Publishers, Leiden/Boston.

Robinson, M. (2002) The right to food: Achievements and challenges. Report by: United Nations High Commissioner for Human Rights. World Food Summit: *five years later*. Rome, Italy, 10 June.

Sabatini, F. (2005). Programming with a human rights approach: A UNICEF experience in operational practice. Chapter 10 in Barth Eide, W. and Kracht, U. eds. Food and human rights in development. Volume I: Legal and institutional dimensions and selected topics. Intersentia, Antwerp and Oxford.

UN (1989). The right to adequate food as a human right. Special Report. UN Study in Human Rights, No 1. Geneva and New York.

Chapter 7

A 'rosy picture'?

Dutch ICESCR report reviewed by UN Committee[228]

Fons Coomans

1. Introduction

In November 2006 the UN Committee on Economic, Social and Cultural Rights reviewed the third Dutch periodic report on the implementation of the International Covenant on Economic, Social and Cultural Rights (ICESCR). The review was completed in dialogue with a delegation from the Dutch government under the leadership of the Dutch Human Rights Ambassador. This contribution discusses and analyses the reporting procedure, whereby attention is also paid to the interaction with NGOs including the NJCM (the Dutch section of the International Commission of Jurists), the position of the Committee and the government, and the Concluding Observations adopted by the Committee after studying and discussing the country report. The Concluding Observations are intended to form the basis for a follow-up process whereby the Committee's recommendations serve as a guideline for (new) legislation and policy. Some suggestions with regard to the substance of this follow-up process are advanced and the question of whether any practical implementation has taken place is discussed. The contribution concludes with a brief look at some recent developments regarding the Dutch government's acknowledgement of the right to water.

2. Objectives of the reporting procedure[229]

The main purpose of a country report is to report to the supervisory Committee on the state of affairs with regard to the realisation of the rights in a country that is party to the Covenant. The country report is part of the reporting cycle whereby a country that is a State party to the Covenant is assessed with regard to the progress, the bottlenecks and problems and any relapses in the realisation of the rights contained in the Covenant.[230] The obligation to report also has another objective, that of conducting a periodic review of national legislation, policy and practice in the light of the obligations contained in the Covenant, partly in relation

[228] An earlier version of this contribution in the Dutch language appeared in the *NJCM Bulletin* (2007: 745-753). The article was updated for the publication in this book whereby particular focus was on the subsequent Dutch periodic government report that was published in the spring of 2008.

[229] On the objectives of the reporting procedure see General Comment no. 1 (1989) of the UN Committee on Economic, Social and Cultural Rights, which can be found in the United Nations Treaty Body Database, www.unhchr.ch/tbs/doc.nsf.

[230] See Article 16 and 17 ICESCR.

to previous Concluding Observations and the General Comments adopted by the Committee. In other words it is a sort of internal evaluation by the government, looking into a mirror, whereby the obligations contained in the Covenant form a framework for review. What effects does the adopted policy have in terms of the rights contained in the Covenant? Is there a willingness on the side of the government to admit that in addition to the positive aspects in this field there are also problems endangering the realisation of the rights? Further to this, an open and honest country report should also encourage public debate in the country that is party to the Covenant on the realisation of the social, economic and cultural human rights.

3. Background to the Dutch report

The third Dutch report was submitted to the UN on 18 August 2005.[231] The official deadline for submitting the report had however expired on 30 June 1997, making the Netherlands about eight years late.[232] No official explanation was offered for this. The report only related to the Dutch part of the Kingdom, thus not to Aruba and the Antilles.[233] Usually it takes some time before a report is put on the Committee's agenda, but in the Netherlands' case the duration was relatively short. This is probably because the Committee did not want to waste any more time. As early as November 2005 a working group of the Committee drew up the List of Issues to be answered by the government in writing prior to the oral discussion.[234] The Committee's Rapporteur for the Dutch report was Mr Sadi, from Jordan. At the meeting the working group gave NGOs the opportunity to make oral observations and provide alternative written information, the so-called shadow reports or parallel reports. The NJCM and the Johannes Wier Foundation (JWS) took this opportunity and in October 2005 they presented a shadow report that was also orally explained during the sitting of the working group on 28 November 2005.[235] Some of the points contained in the List of Issues seem to be based directly on information submitted by the NGOs. An example of this is the Committee's enquiry regarding the manner in which the government adopts a human rights approach in the field of development cooperation and the manner in which the government's stance on economic and social rights affects this cooperation.[236]

[231] UN Doc. E/1994/104/Add.30 (23 August 2005).

[232] In fact the fourth periodic report was due in June 2002. On the Netherland's failure to submit reports on time to the UN Convention Committees see *METRO*, 29 March 2007, p. 1 and 5, and the government's response to questions from Van Dam, MP (PvdA), *Aanhangsel Handelingen II* 2006-2007, no. 1298.

[233] The third report from the Dutch Antilles, UN Doc. E/C.12/ANT/3, was reviewed by the Committee in May of this year. See the report on the review www.unhchr.ch/tbs/doc.nsf.

[234] UN Doc. E/C.12/Q/NLD/3 (2 December 2005).

[235] Contribution of the Dutch Section of the International Commission of Jurists (NJCM) and the Johannes Wier Stichting to the Committee on Economic, Social and Cultural Rights, 16 pages, October 2005, accessible at: www.njcm.nl.

[236] See UN Doc. E/C.12/Q/NLD/3, § 4 and shadow report NJCM and JWS, p. 3.

The government gave a fairly detailed response to the written questions of the Committee in July 2006.[237] The government invited the NJCM and the JWS to The Hague, in preparation for the review of the Dutch report in Geneva, to become better acquainted with the officials who were to represent the Netherlands and to exchange ideas about the objectives, procedures and follow-up of the Covenant reports and the content of the ICESCR report. Both NGOs accepted this invitation subject to the explicit notation that such a meeting may not be considered to be a consultation round with NGOs, in other words each party retained its own role and responsibilities in the reporting procedure. The Dutch human rights ambassador headed the delegation. The Ministries of Foreign Affairs, Justice, Home Affairs, Social Affairs and Employment, Health Welfare and Sport and the Permanent Representative of the Netherlands in Geneva also formed part of the delegation. The Departments of Housing, Environmental Planning and Education, Culture and Science were not represented in the delegation.

4. Features of the Dutch government report

The Dutch report, 155 pages long excluding annexes, covered the period running from 1 January 1996 to 31 December 2002. This meant that when the report was reviewed by the Committee in November 2006 the information contained therein was to a certain extent out of date. The government had recently submitted supplementary information in its written response to the List of Issues and also did so during the oral presentation of the report in Geneva. The report primarily adhered to the guidelines for drawing up country reports as laid down by the Committee.[238] The report contained detailed information about legislation and policy relating to the rights contained in the ICESCR, although this was mainly descriptive in nature. It dealt with no or hardly any problems or bottlenecks in the realisation of the rights in the Netherlands or policy effects in terms of rights. This gave it a somewhat rosy colour, whereas it is generally known that the situation in the Netherlands with regard to these rights is not equally positive for everyone. Consider for instance the access to health care of vulnerable groups such as immigrants residing unlawfully in the Netherlands.

One section of the report that can be regarded favourably is the detailed manner in which the government responded to the previous Concluding Observations made by the Committee in 1998.[239] A clear omission in the report is any sort of reference to Dutch case law on the rights contained in the Covenant. An appeal to provisions in the ICESCR or to the European Social Charter is almost always rejected by the Dutch courts on the grounds that these are not directly

[237] UN Doc. E/C.12/NLD/Q/3/Add.1 (17 July 2006).
[238] UN Doc. E/C.12/1991/1.
[239] UN Doc. E/C.12/1/Add.25 (16 June 1998). With respect to this see F. Coomans, 'Nederland voor het VN-Comité inzake economische, sociale en culturele rechten', *NJCM Bulletin* 1998, p. 939-946.

applicable.[240] This information would have been important for the Committee in enabling it to better understand why the rights have no direct significance for citizens in the Netherlands.

5. The dialogue between Committee and government

An important issue was the status of the ICESCR in the legal order of the Netherlands. In 1998 the Committee had recommended that the government reconsider its rejection of the ICESCR provisions as being directly applicable.[241] However, in its third report the government stuck to the familiar point of view rejecting the direct applicability of provisions 'Simply accepting direct application of provisions that need to be worked out in greater detail would mean that it would be left to the national courts to put the objectives set down in the Covenant into practice'. That would be contrary to the principles of a democratic decision-making process, as this operates in the Netherlands.[242] In its shadow report, on this issue one of the questions asked by the NJCM was whether the government agreed with the Committee's opinion that some elements of the rights contained in the Covenant are directly applicable in many countries that are party to the Covenant.[243] According to the Human Rights Ambassador the implementation of the human rights contained in the Covenant is 'essentially a political question', meaning the courts have no role to play here. On being asked what its position was with regard to the creation of an Optional Protocol to the ICESCR allowing an individual to make a complaint against a state, the government replied that the Netherlands' approach was cautious but constructive. In connection with this he did acknowledge there to be a contradiction in the Dutch stance that no distinction can be made between civil and political rights and economic, social and cultural rights, but that different mechanisms can apply to the implementation of these rights.[244]

The issue of access to health care for immigrants residing illegally in the Netherlands was only referred to indirectly. The government had made no mention of this in the country report. The JWS had drawn attention to this in the shadow report.[245] Only one member of the Committee asked a question about this. The government replied that immigrants residing illegally in the Netherlands have no access to

[240] See CRvB 25 May 2004, LJN: AP0561 and CRvB 18 June 2004, *AB* 2004/296; CRvB 2 December 2003, *USZ* 2004/88; HR 14 April 1989, *NJ* 1989, 469 and HR 7 May 1993, *AB* 1993, 440; ABRvS 6 March 2002, *Gemeentestem* 2003, 62; Rb. Den Haag, 7 July 2006, LJN: AY1710.
[241] UN Doc. E/C.12/1/Add.25, § 21 and Coomans, supra footnote 239, p. 942.
[242] UN Doc. E/1994/104/Add.30, § 8.
[243] Shadow report NJCM and JWS, p. 2.
[244] UN Doc. E/C.12/2006/SR.33, § 47.
[245] Shadow report NJCM and JWS, p. 14.

health care except for emergency assistance, which includes anti-AIDS treatment.[246] This issue was not referred to in the Concluding Observations.

A major subject of discussion was prostitution in the Netherlands. In the List of Issues the Committee had suggested a link between the legalisation of prostitution and trafficking in people for the purposes of sexual exploitation. The government denied such a link. The legalisation of brothels makes it easier to combat the exploitation of women.[247] However, some members of the Committee continued to argue that there is an 'obvious link between prostitution and trafficking'.[248] According to the government the Committee based its arguments on outdated statistics, in other words pre-2000, before prostitution was made legal. The government claimed that recent research conducted by a number of NGOs indicates the number of sex establishments to be decreasing.[249] The Committee did not ask for more details about up-to-date and exact statistics.

Another issue that was dealt with was the (sexual) abuse of children, including child pornography on the Internet. These problems were also touched upon in the government report, whereby it was stated that the National Action Plan to Combat the Sexual Abuse of Children (NAPS) had reached completion in late 2002.[250] The List of Issues contained a query as to why the Action Plan had not been extended. The government replied that the positive results of this Action Plan are now part of regular policy. This included, for instance, the setting up of a reporting procedure for specific occupational groups in cases where child abuse is suspected.[251] The scope of the problem of child abuse was neither stated in the government report nor in the response to the List of Issues. This was looked into during the oral review and the government indicated that approximately 100,000 children are the victim of child abuse each year.[252] In the Concluding Observations the Committee expressed its concern about this evil.[253]

The issue of domestic violence was prominent during the review in Geneva. The government report described the scope of the problem, the policy objectives and measures that have been taken and are yet to be introduced.[254] As regards the scope, in the report the government stated that from a survey conducted amongst

[246] UN Doc. E/C.12/2006/SR.34. This document containing the report of part of the review of the Dutch report was not available at the time this contribution was written. This footnote is therefore based on my own notes made during the meeting which I attended as an observer.
[247] UN Doc. E/C.12/NLD/Q/3/Add.1, p. 18; E/C.12/2006/SR.35, § 7, 29.
[248] UN Doc. E/C.12/2006/SR.35, § 24.
[249] UN Doc. E/C.12/2006/SR.35, § 32. The government omitted to state exactly which research this referred to.
[250] UN Doc. E/1994/104/Add.30, § 313.
[251] UN Doc. E/C.12/NLD/Q/3/Add.1, p. 17-18.
[252] UN Doc. E/C.12/2006/SR.34.
[253] UN Doc. E/C.12/NLD/CO/3, § 16.
[254] UN Doc. E/1994/104/Add.30, § 85-97.

both men and women it appears that over 40% of this group of respondents has at some time been the victim of some form of domestic violence.[255] The List of Issues inquired into the reasons behind this high percentage. However, the government's reply did not contain an explanation but was rather a more detailed description of the scope of the problem.[256] A number of Committee members were interested in the question of why the Netherlands has no separate legislation on domestic violence.[257] The government replied that although there is indeed no specific legislation to deal with this abusive situation, the use of violence is a criminal offence pursuant to the Penal Code.[258]

6. The Concluding Observations

The Concluding Observations comprise three parts: positive aspects, points for concern and suggestions and recommendations. With respect to positive aspects the Committee named the measures taken to encourage elderly people to remain active in the labour market. Furthermore the Committee welcomed the abolition in 2005 of school fees and study fees for 16- and 17-year-olds in secondary education. Finally, the Committee is positive about the human rights approach upon which the Dutch policy on development cooperation is based.[259]

Under points for concern the Committee names the problems with respect to the direct applicability of the ICESCR provisions, the low level of labour market participation of women from ethnic minorities, the trafficking in people for the purpose of sexual abuse, child abuse and the continuation of domestic violence and lack of specific legislation to combat this abusive situation.[260]

Under the heading suggestions and recommendations the Committee lists an entire series of proposals that the government could introduce to strengthen the implementation of the ICESCR in the Netherlands. First and foremost the Committee encourages the government to give its support to the process of creating an Optional Protocol to the ICESCR allowing an individual to make a complaint against a state.[261] Furthermore, the Committee recommends that the Netherlands, as a party to the Covenant, review the degree to which ICESCR provisions can be directly applied. In connection with this the Committee puts pressure on the country 'to ensure that the provisions of the Covenant are given effect by its domestic courts, (...), and that it promotes the use of the Covenant as a domestic

[255] Idem, § 90.
[256] UN Doc. E/C.12/NLD/Q/3/Add.1, p. 15.
[257] UN Doc. E/C.12/2006/SR.34.
[258] UN Doc. E/C.12/2006/SR.35, § 39.
[259] UN Doc. E/C.12/NLD/CO/3, § 6-8.
[260] UN Doc. E/C.12/NLD/CO/3, § 11, 13-16.
[261] UN Doc. E/C.12/NLD/CO/3, § 17.

source of law'.[262] It is interesting to note here that this recommendation is addressed to the Netherlands as a country that is party to the Covenant, therefore also to the three powers that perform the most important functions of the country, the legislator, executive and judiciary. All this means that all three have their own responsibilities with regard to the application of ICESCR standards.

It is also recommended that the labour market participation of women from ethnic minorities be increased.[263] It is not surprising that recommendations were also made to tackle the sexual abuse of women and children and child pornography. The Committee also emphasises the need to introduce specific legislation on domestic violence.[264] Finally the Committee requests that the Netherlands submit the following fourth and fifth country reports combined before 30 June 2007.[265]

7. Evaluation of the dialogue between the Netherlands and the Committee

Looking back at the dialogue it can be generally concluded that this was satisfactory. The Committee's chairman referred to the dialogue as being 'fruitful and constructive'.[266] The oral explanations given by the members of the Dutch delegation were extensive, fairly detailed and presented in an open manner. Contrary to the Dutch report, in the oral review the bottlenecks were acknowledged and recent data was provided. The latter being all the more important because the report only covered the period up to the end of 2002. The Committee shared the opinion that the dialogue with the Dutch government was open and constructive.[267] There was a good division of work within the Dutch delegation, in which the members also complemented one another. However, I am of the opinion that the Dutch government laid too much emphasis on intended policy and too little on the results and effects of previous policy in terms of the rights contained in the Covenant. An example with respect to this is the effects of the privatization of the social security system on the right of everyone to equal access to social security facilities.

The questions asked by the Committee members were generally critical and reflected an attentive study of the Dutch situation. A considerable amount of emphasis was given to subjects with respect to which the Netherlands played a progressive role in the past, such as the legalisation of prostitution, as well as to sensitive and difficult subjects such as domestic violence and sexual exploitation and the abuse of women and children. These are subjects whereby opinions can

[262] UN Doc. E/C.12/NLD/CO/3, § 19.
[263] UN Doc. E/C.12/NLD/CO/3, § 22-23.
[264] UN Doc. E/C.12/NLD/CO/3, § 26-28.
[265] UN Doc. E/C.12/NLD/CO/3, § 34.
[266] UN Doc. E/C.12/2006/SR.35, § 41.
[267] UN Doc. E/C.12/NLD/CO/3, § 3.

differ in accordance with principles. These subjects were raised by the rapporteur of the Committee for the Dutch report in particular. Sometimes the members of the Committee did not pursue their questions and the government got away with a general answer that hardly added anything to the information contained in the government report or in the written replies. Examples of this are the government's replies regarding the direct applicability of the ICESCR in the Dutch legal order and the position with regard to a Protocol to the ICESCR allowing an individual to make a complaint against a state.

In my opinion the Concluding Observations are a little on the thin side due to the omission of a number of major issues such as the review of the healthcare system, the access to healthcare for immigrants residing illegally in the Netherlands and the issue of the 'white' and 'black' schools. These issues were referred to in the shadow reports of the NJCM and the JWS, but were not fully taken up by the Committee. The reasons for this could be that in the Committee's view the government gave a convincing response during the oral examination of the report or in the written reply to the List of Issues, or that the Committee had made a selection of issues and considered the problems in question to be of less importance. Of course it could also be the case that these issues escaped the Committee's attention amidst the other subjects.

In the previous Concluding Observations dating from 1998 the Committee remarked that 'the Netherlands has to a considerable extent met its obligations with respect to the protection of the rights set out in the Covenant'.[268] Such an evaluation is lacking in the Concluding Observations dating from 2006. During the review of the report the rapporteur of the Committee observed that the Dutch delegation had outlined a rosy picture of the Dutch achievements in the field of realising the social, economic and cultural rights. He found it hard to believe that there were no 'challenges, crises or problems'.[269] The Human Rights ambassador replied to this by stating that it was not the intention to sketch an 'overly positive picture' of the situation in the Netherlands in the report. There were obviously challenges to be met.[270] This is an acknowledgement of the bottlenecks pointed out by the Committee in the Concluding Observations. The extent to which economic, social and cultural rights have been realised in the Netherlands is, especially in comparison with other countries, significant, yet there are some structural problems that unfortunately were not given the attention they deserved during the review of the Dutch report. For instance, the integration of migrants in the labour market and the rights of asylum seekers and immigrants residing illegally in the Netherlands. These problems were addressed, but only superficially.

---

[268] UN Doc. E/C.12/1/Add.25, § 4.
[269] UN Doc. E/C.12/2006/SR.33, § 27.
[270] UN Doc. E/C.12/2006/SR.33, § 45.

8. Follow-up and evaluation

The Committee asked the government to give the Concluding Observations wide publicity and to stipulate in the next report the measures adopted for putting the recommendations into practice. The government was also called upon to enter into dialogue with NGOs prior to submitting the next report.[271] In the dialogue with some NGOs in October 2006 the Human Rights Ambassador had already announced a willingness to talk to NGOs with regard to the potential implications of the Concluding Observations for the Dutch situation. This is a positive development. The ICESCR is not well known in the Netherlands as a normative treaty. Consequently, for example, the Dutch press was not represented at the review of the Dutch report in Geneva and the report was given no attention by the Dutch newspapers. There are only a few NGOs that draw attention to it and test the government's policy against the ICESCR standards. The trade unions form no exception to this. Neither is the ICESCR's existence truly recognised in parliament. Only a small number of members of parliament are informed and attentive. [272] There is no real public debate regarding the implications of the ICESCR for the Netherlands. A follow-up meeting attended by a wide range of participants from non-governmental organisations, including occupational groups and social organisations, parliament and the judiciary and universities, as well as representatives from the Committee could increase familiarity with the ICESCR standards in Dutch society.[273] A National Human Rights Institute could play a leading role here. However, in my opinion the most important obstacle remains the fact that the ICESCR provisions are not directly applicable in the Netherlands. They are not much use to citizens; they cannot be relied upon with any reasonable chance of success. As a result of this the position of the ICESCR in the Netherlands is weak. The Concluding Observations call upon the government and other government agencies to break through this situation. The Concluding Observations also urge the Dutch government to give a different signal with regard to the carry-over effect of the ICESCR provisions. But the judiciary and members of parliament can, each in his or her own way, also prompt the government to abandon its traditional point of view. The negotiations regarding

[271] UN Doc. E/C.12/NLD/CO/3, § 32. As already stated the date for submitting the combined fourth and fifth report is 30 June 2007. This report is to cover the period 2003-2006. We must wait and see whether the report is actually on time. See also the Letter from the Minister for Foreign Affairs of 2 February 2007, *Kamerstukken II* 2006-2007, 30 800 V, no. 62.

[272] For example, see the questions asked by Peters, member of the Lower Chamber (Groen Links) about the promises made by the Netherlands when standing for a seat in the UN Human Rights Council, *Aanhangsel Kamerstukken II* 2006-2007, no. 1352.

[273] Germany has experience of these types of meetings. See the report by the German Human Rights Institute, *Examination of State Reporting by Human Rights Treaty Bodies: An Example of Follow-Up at the National Level by National Human Rights Institutions*, April 2005, available at www.institut-fuer-menschenrechte.de. Further to the previous ICESCR Concluding Observations the NJCM organised a follow-up meeting with representatives from departments, NGOs and universities. See, *NJCM Bulletin* 2000, p. 1134.

an Optional Protocol to the ICESCR allowing individuals to make a complaint against a state[274] provide an ideal opportunity for the government to give such positive signals indicating a change of tack.

9. The new Dutch report in 2008

In the spring of 2008, thus later than the date stipulated by the Committee of 30 June 2007, the government published the combined fourth and fifth report covering the period between January 2003 through December 2006.[275] This report consists of two parts. The first part contains the government's response to the Concluding Observations of November 2006. The second part reports on the implementation of the Articles 2 and 6-15 of the Covenant. As regards the publicity given to the Concluding Observations the report states that these have been primarily distributed amongst civil servants and members of parliament.[276] The report makes no mention of a dialogue with or the involvement of NGOs and civil society in the preparation of the new government report. This is disappointing. As far as I am aware the Concluding Observations have not been debated in parliament. Unfortunately in the report the government fails to go into the Committee's urgent request that the position of the ICESCR in the Dutch legal order be promoted to that of a source of law. The report only states that 'the Netherlands would refer to previous reports for information on the Netherlands' position on the direct applicability of the Covenant'.[277] From this it can be deduced that for the time being there is no change in the Dutch position regarding this issue. This does not detract from the fact that the government should have gone into more detail with respect to this issue.

As regards the state of affairs relating to the establishment of an Optional Protocol to the ICESCR allowing individuals to make a complaint against a state, a draft text has been adopted by the UN Human Rights Council on 18 June 2008. The text is now to be submitted to the General Assembly for final acceptance.[278] This will probably take place in December 2008. The Netherlands, as a member of the Human Rights Council, has not hindered the text's progression to the General Assembly. According to the government agreement in the Council regarding the text's progression to the General Assembly does not imply agreement on the content of the Protocol, but an acknowledgement of the fact that the text is the best possible result given the relationships in the Working group that negotiated

[274] The text of a draft Protocol is published as UN Doc. A/HRC/6/WG.4/2 (23 April 2007).
[275] UN Doc. E/C.12/NLD/5.
[276] UN Doc. E/C.12/NLD/5, § 121.
[277] UN Doc. E/C.12/NLD/5, § 11.
[278] The draft text of the Protocol is contained in UN Doc. A/HRC/8/7, Annex I.

the text.[279] In other words the Netherlands reserves its opinion with regard to the Complaints Protocol.

As regards the report on the right to food the Dutch 2008 report is downright disappointing. A substantial part is identical to the 2005 report on the right to food. There is a single sentence that refers to the emergence of Food Banks in the Netherlands: 'Recently private initiatives have been launched to further assist people dependent on social security by providing low-priced food to specific target groups'.[280] However, nothing is said about the background to this emergence of Food Banks and why they are apparently providing for a need.

Under the heading 'Equitable distribution of world food supplies' there is a brief explanation of the objectives of the Dutch development policy.[281] Indeed here it is stated that the Dutch government's policy is intended to tackle the food insecurity of vulnerable groups, but there is no vision on this from the perspective of the human rights approach.

10. Other developments

Another important, positive development is that, through Minister Verhagen in a speech for the UN Human Rights Council, on 3 March 2008, the Dutch government recognized the right to water and sanitation.[282] According to the government this recognition has no legal consequences because the right to water and sanitation is already embodied in the right to an adequate standard of living and the right to the highest attainable standard of health contained in the ICESCR (Article 11 and 12), which the Netherlands has already recognised and accepted. The government does, however, believe that recognition has a political significance. This significance shall primarily be expressed in the policy dialogue with partner states within the scope of development cooperation. Recognition of this right allows the Netherlands to legitimately point out the obligations of governments in partner states and the rights of the population to adequate basic facilities.[283] The government is also to place the right to water on the agenda of international financial institutions, such as the IMF and the World Bank. This means, among other things, that the Netherlands will insist that the human rights perspective be taken into account when preparing decisions regarding major water and sanitation projects.[284]

---

[279] See the government's replies to questions put by Van Dam, Member of Parliament, on the Netherland's position with respect to the Protocol to the ICESCR allowing individuals to make a complaint against a state, *Aanhangsel Kamerstukken II* 2007-2008, no. 2015.
[280] UN Doc. E/C.12/NLD/5, § 219.
[281] UN Doc. E/C.12/NLD/5, § 228-232.
[282] Policy memorandum Development Cooperation, *Kamerstukken II* 2007-2008, 31 250, no. 15, p. 1. See also, W. Thiebou, 'Water als mensenrecht', *VN Forum* (Journal of the Dutch Association for the United Nations) 2008-3, p. 16-22.
[283] Policy memorandum Development Cooperation, p. 1, 3.
[284] Policy memorandum Development Cooperation, p. 3.

The government considers the significance of the right to water to lie primarily within the scope of development cooperation and thus to have most relevance for countries in the South. This is obviously correct insofar as countries where water is not yet an accessible and affordable basic facility for everyone concerned. However the right to water as a human right can also be significant for the Netherlands itself. This was apparent when, with reference to Minister Verhagen's speech for the Human Rights Council and the ICESCR, the subdistrict court at Heerlen ruled that disconnecting the water supply is not permissible in certain circumstances. In the case in question this applied because the defendant was dependent on the Water Supply Company Limburg that should be classed as a regional monopolist. Furthermore, disconnection was not proportionate to the payment arrears of the defendant.[285] With reference to this ruling Minister Cramer of VROM (Ministry of Housing, Spatial Planning and the Environment) was quick to remark in a letter to parliament that the right to water did not imply that the water should be supplied free of charge and that as a last resort water supply companies can disconnect defaulters from the water supply. They do not consider this to be contrary to the right to water.[286]

[285] Subdistrict court Heerlen, 25 June 2008, *LJN: BD 5759.*
[286] Policy memorandum Development Cooperation, *Kamerstukken II* 2007-2008, 31 250, no. 33.

Chapter 8
Regulating biofuels in the name of sustainability or the right to food?
The case of an emerging policy field in the Netherlands

Otto Hospes

1. Introduction

It is now widely acknowledged that the increased use of food crops for making energy was one of the causes of the global food crisis in the years 2007-2008, next to severe droughts, low stocks, high prices of fossil fuels, increasing food demands, underinvestment in agriculture and intensified export restrictions (OECD, 2007; OECD and FAO, 2007; Senauer, 2008; Von Braun, 2007; World Bank, 2007). According to former UN secretary-general Kofi Annan, people in more than 40 countries have taken to the streets to protest against the increase in food prices and to demand their right to food. But who is and can be held responsible for their deteriorating food situation if the causes of the food crisis are so diverse, complex and interdependent? Sen postulated that democracies and free press function as safety valves that prevent food crisis turning into famines because democratically elected politicians are sensitive to information on the deteriorating food situation of potential voters (D'Souza, 1994; Sen, 1981). But does this mean that we can safely assume that the breaking news on protests in more than 40 countries against high food prices will soften the effects of the current food crisis?

No, it doesn't. The food crisis is not just a national and technical affair, but foremost a dramatic example of governance failure and complexity in a new global era. Many state and non-state actors agree on the complex causes of the dramatic rises of food prices in the years 2007-2008 but at the same time hardly consider the full implications of the failure of states to address this crisis – either individually or collectively as members of international organisations. Governments define key issues and propose general principles or technical solutions to address the current food crisis (like investing more in agriculture and in second-generation biofuel technologies) but only a few call for a redefinition of responsibilities of state and non-state actors, new institutional arrangements and modes of governance at the same time.

The Dutch government has an international reputation as a primary mover in developing biofuel policies and very much wants biofuel production not to compete with food production. This article focuses on a testing framework for sustainable biomass that has been adopted by the Dutch government. The central questions are: to what extent is the right to food central in this framework? To what extent is the Netherlands' quest for sustainability of biomass production defined in terms of state obligations and rights of individuals? Does it make a difference whether sustainability or the right to food is taken as the point of departure to regulate biofuel production?

To give an answer to these questions and to put the Netherlands' testing framework into an international perspective, I will first present and analyse two UN frameworks on bioenergy or biofuels. My aim is not only to show to what extent the Dutch framework is unique but also to explain the implications of taking – or not taking – the right to food as the point of departure for developing biofuel policies. For this purpose, I have focused on two authoritative reports in which the FAO took a leading role. I have selected the FAO writings because the FAO is the UN agency where the worldwide exchange of ideas and findings on the relationship between biofuel production and food security is most important and comprehensive. Of course, many other multilateral agencies and international organisations (like OECD, World Bank, IFPRI) have produced reports about biofuels but none of these considers food security as its mission statement.

2. Two different UN frameworks for national governments

In 2007 the UN-Energy group published 'Sustainable Bioenergy: A Framework for Decision Makers'. Some 20 UN organisations and programs constitute the membership of UN-Energy that presents itself as the 'principle interagency mechanism in the field of energy'. Its two objectives are 'to ensure coherence in the UN system's multi-disciplinary response to the World Summit on Sustainable Development (WSSD) and collective engagement on non-UN stakeholders.' The ambition level of the framework is rather modest: it is 'not designed to provide prescriptive measures but rather to identify areas that require priority attention at the national and international levels.' The report was sponsored by the FAO, which took a leading role in drafting the document, drawing on important support from the Worldwatch Institute.

The bottom line of the report is that bioenergy offers both opportunities and threats in terms of poverty alleviation, job creation, agricultural development, food security, energy security, biodiversity and climate change. The main message of the report is two-fold. First, the authors advise decision makers to think twice, not to rush. They recommend that the economic, environmental and social impacts of bioenergy development must be assessed before deciding if and how rapidly to develop the industry and what technologies, policies and investment strategies to pursue. Secondly, the authors recommend incorporating social and environmental concerns from the very start of policy making as they fear that, 'Unless new policies are enacted to protect threatened lands, secure socially acceptable land use, and steer bioenergy development in a sustainable direction overall, the environmental and social damage could in some cases outweigh the benefits' (UN-Energy, 2007: 5).

Nine key sustainability issues are distinguished in the report. One of these issues is about the implications for food security: to understand key relationships between expansion of liquid biofuel production and food security, the authors first propose to develop an analytical framework to understand country-specific effects of expansion of liquid biofuel production on food security in terms of availability, access, stability and utilisation. Secondly, they propose investing more in agricultural research aimed at the development of energy cropping that would not compete directly with food production. Thirdly, they propose and call on policymakers to better understand the policy nexus for liquid biofuels:

> *At least four distinct policy domains are shaping development of the liquid biofuels sector: energy, environment, agriculture and trade. Similarly, policies at the national, regional and global levels are highly relevant and may interact in unexpected ways. Policy makers need to understand the interactions among these various policy domains and levels to ensure that food security considerations are given priority (UN-Energy, 2007: 35).*

In my view, all three proposals to assess implications of bioenergy for food security make sense but are very much technical recommendations, emphasising the need for analysis, research and awareness building. A major shortcoming of the 'Framework for Decision Makers' is that the term 'decision makers' is neither described in a general way nor in terms of responsibilities, obligations, competences. None of the proposals is about who or what institution should take the lead in developing an analytical framework, new agricultural research or insights on the policy nexus for liquid biofuels. Considering the membership of the UN-Energy group, one could assume that decision makers are UN agencies or member governments. But what about multinational companies, NGOs or public-private partnerships? None of the proposals deals with the question about who is responsible for the effects of bioenergy development behind and beyond national borders: do national governments, for instance, have extra-territorial obligations? The proposal to understand the policy nexus remains limited to understanding interactions among various policy domains. It does not raise the difficult question on how to develop biofuel policies if and when policy makers from different domains and at different levels consider biofuel as one of their competences. No consideration is given to public-private interactions that play or could play an important role in shaping public policies on sustainable bioenergy. Sustainability of bioenergy is the key concern but the issue of governing sustainability is ignored or simplified as a problem of reducing lack of knowledge and creating awareness about bioenergy as an interdepartmental policy nexus.

One of the few UN reports that does take up governance issues is the report of Asbjørn Eide on 'The Right to Food and the Impact of Liquid Biofuels (Agrofuels)'. It was published in a series of Right to Food Studies of the FAO in 2008, that is, only one year later than the UN-Energy report. Like the UN-Energy report, written under the stewardship of FAO officer Gustavo Best, the report of Asbjørn Eide was also funded by the FAO. Remarkably, the Right to Food study explicitly mentions that 'positions and opinions presented do not necessarily represent the views of the FAO' whereas the UN-Energy report that was sponsored and written by the FAO, lacks such a clause.

The contents of the two reports differ from each other in at least three ways. First, in the report by Eide the impact of biofuels is assessed from the perspective of food security only, not from a broad sustainability perspective. Second, his report is not about possible benefits and threats of biofuel expansion but firmly concludes that liquid biofuel production has indeed helped to weaken the access to adequate food or to the resources with which vulnerable people can feed themselves, in at least three ways: (a) by contributing significantly to the increase in food prices; (b) by causing land concentration for plantation-type production; and (c) by causing a number of environmental problems, reducing biodiversity and leading to competition for water. The third and possibly most fundamental difference with the report of the UN-Energy group is that the Right to Food study is normative and prescriptive. Eide considers the human right to food as the framework for the development of biofuel policies and projects. According to Eide, 'the right to food has to be at the foundation of efforts to handle the current world crisis caused by soaring food prices' (Eide, 2008: 25). In contrast to the framework proposed by the UN-Energy group, the right to food is not an *analytical* framework directed at the understanding of complex relationships between biofuel expansion and food security but a *normative* framework to assess and to develop biofuel policies in terms of human rights and state obligations. In a systematic way, Eide uses the right to food to address the ethical dilemma of food versus fuel, to highlight state obligations at the national and international level, and to propose institutional arrangements.

Taking the right to food as the foundation for assessing the impact of biofuels on food security, Eide declares the issue of food versus fuel as no ethical dilemma at all. He simply states that the right to food has 'an obvious priority over biofuel'. In his view, 'much of the energy consumption particularly in developed countries is unnecessary by a wide mark' (Eide, 2008: 26). Though he considers the increasing demand for meat and fish for the growing numbers of middle-class consumers in countries like China and India as one of the general and long-lasting causes of increase in food prices, Eide limits his analysis to the impact of liquid biofuels. His argument is that effects of increasing meat consumption are probably impossible to mitigate as this situation is likely to continue. He thinks that biofuel production in competition with food, however, can be reduced 'because it depends on political decisions' (Eide, 2008: 15).

In the view of Eide, national governments, particularly the EU and USA, have created an artificial market for biofuel: they have set targets on biofuel consumption, prescribed mandatory blending and/or provided tax breaks and subsidies to companies to invest in biofuel production. These are political decisions. Eide, however, does not start by saying that national governments need to review these decisions but, more fundamentally, first reminds them of their own promises, laid down in international human rights law, to respect, protect and facilitate the human right to food (see Chapter 2 by Van der Meulen and Vlemminx in this volume and Hospes, 2008a; 2008b).

The right to food was recognised in 1948 as part of the Universal Declaration on Human Rights. The right to food is the right of every human being to be free from hunger and to feed him or herself with nutritionally adequate and culturally acceptable food. By signing the International Covenant of Economic, Social and Cultural Rights (1966), 157 states have imposed three obligations on themselves in relation to the right to food:

- the obligation to respect: to do no harm to existing access to food;
- the obligation to protect: to ensure that enterprises and individuals do not deprive individuals of their access to food;
- the obligation to facilitate: to create an enabling, non-discriminatory policy environment and to build food stocks and social safety nets for those who cannot feed themselves (Hospes, 2008a; 2008b).

According to Eide, the first obligation implies that in contemplating projects for biofuel production, states should respect existing access by rural people of their established sources of livelihood and should abstain from measures of eviction and resettlement. Respecting existing access to food also requires that states should not adopt policies that deliberately raise the prices of basic staples without taking measures to shield food insecure people from its impact. The second obligation implies that the state must protect peoples' livelihood against third parties, and against impermissible initiatives by investors to take over land for biofuel production. Existing land rights or user rights for the existing rural population should not be threatened by biofuel plantations and other projects. This very much resembles the proposal of the UN-Energy group to enact 'new policies to protect threatened land and to secure socially acceptable land use', with a view to preventing social damage by biofuel policies from outweighing the benefits (UN Energy, 2007: 5). However, unlike the UN-Energy group, Eide considers such measures not to be a policy option but a state obligation. The third obligation is to establish safety nets to take care of those who are negatively affected by biofuel projects. A cost-benefit analysis of any biofuel project should include the costs of the safety net.

It is the self-imposition and mutual agreement of states with these three obligations which makes Eide believe that a 180-degree turnaround of political commitments to biofuel targets has a firm legal ground and is not just ethics. This turnaround is a process for which the body of international law on the right to food (see Hospes, 2008a) not only provides a substantive direction but also a research focus and an institutional agenda.

One of the first steps proposed by Eide to national governments is to conduct research prior to the adoption and implementation of biofuel policies. This very much resembles a key message of the UN-Energy group that warns against rushing into bioenergy and calls for an impact assessment before developing the industry. In contrast to the UN-Energy group, however, Eide puts a strong emphasis on statistical disaggregation and explaining why particular groups have become more food insecure than others. He particularly calls on states to establish and keep updated Food Insecurity and Vulnerability Mapping Systems (FIVIMS) and to learn from experiences in other countries of smallholders, workers, contract farmers and/or indigenous peoples, who were involved in the establishment or expansion of biofuel plantations. More fundamentally, Eide does not believe that scientific knowledge or more insight into complex relationships between biofuels and food security are enough to address or prevent possibly the negative effects of biofuels on food security.

Research on how and why which particular group of people will be affected by increased use of land, water and crops for making energy, should be part of a national strategy for the realisation of the right to food. Quoting the UN Rapporteur on the Right to Food, Olivier de Schutter, Eide distinguishes three other process requirements of this national strategy: (a) improving the coordination between different relevant ministries as to mainstream the right to food into other policy domains or sectors (like energy, trade, agriculture); (b) improving accountability, with a clear allocation of responsibilities; and (c) ensuring adequate participation, particularly of the most food insecure segments of the population.

This institutional agenda is an agenda at the national level, as the general rule under human rights law is that each state separately has the primary responsibility to respect and ensure human rights for everyone within its territory; other states and the international community in general have a supplementary role. Eide, however, believes that in the case of the challenges to the right to food posed by biofuels, states do also have 'transborder duties' or 'extra-territorial obligations' (see Chapter 9 by Hausman in this volume). He even proposes to 'reverse the order' as 'problems are often caused by actors having a global reach' (Eide, 2008: 43). Considering that biofuel targets of the EU and the USA can only be met if there is also sizeable import from developing countries, Eide argues that 'importing states have a duty to protect the local population in exporting countries by establishing barriers against import of biofuel which is produced in socially and environmentally harmful ways' (Eide, 2008: 44). To respect the right to food, Eide proposes that these states bring an end to targets, mandatory quota for blending, tax breaks and any other preferential treatment of biofuel that may indirectly lead to land conflicts, land eviction, sharp increases in prices of staple food and other social damage in developing countries or emerging economies, where biofuels are produced and exported.

3. Is the Dutch framework on sustainable biomass sustainable?

In contrast to the development of guidelines on the right to food (see Chapter 6 by Oshaug in this volume), the Netherlands government has taken a very pro-active and progressive role in formulating universal principles and criteria for sustainable biomass. In 2005 the Dutch government gave an assignment to a working group to organise a multi-stakeholder consultation on sustainable biomass. Professor Jacqueline Cramer led the working group. Her name was given to the report by the working group, officially known as the 'Testing Framework for Sustainable Biomass'. In February 2007 she became minister for Housing, Spatial Planning and Environment. Two months later minister Cramer submitted the Cramer report to the Dutch parliament.[287]

The key concept of the Cramer report is 'sustainability', which has been used to bring together different concerns on the environmental, economic and social impact of increased use and production of biomass for energy production. Six themes have been identified to distinguish different dimensions of sustainability of biomass production at the level of the enterprise and at the national or regional level ('macro effects'): greenhouse gas emissions, competition with food and other local applications, biodiversity, environment, prosperity, and social well-being. For every theme, one to three principles have been formulated (see Box 1):

The heart of the report is formed by criteria and indicators for every principle. They are supposed to form the basis for certification of sustainable biomass production, whether produced in or outside the EU. However, for two of the six themes neither criteria nor indicators have been determined yet: competition with food and other local applications, and prosperity. For the first of these themes the principle is that the production of biomass for energy must not endanger the food supply and other local applications. For the second, the principle is that the production of biomass must contribute towards local prosperity (see Box 1).

For both themes, data collection and monitoring is proposed as a first step and minimum requirement. As regards the theme of competition with food, the report calls for insights into the change of land use and change of prices of food and land in the area of the biomass production unit; as regards the theme of prosperity, the advice is to collect information at a company level on direct economic value

[287] Letter to the House of Commons 30305, nr. 25, annex.

Box 1. Six themes and nine principles for the production of sustainable biomass.

Theme 1: greenhouse gas emissions	Principle 1: The greenhouse gas balance of the production chain and application of the biomass must be positive. Principle 2: Biomass production must not be at the expense of important carbon sinks in the vegetation and in the soil.
Theme 2: competition with food and other applications	Principle 3: The production of biomass for energy must not endanger the food supply and local biomass applications (energy supply, medicines, building materials).
Theme 3: biodiversity	Principle 4: Biomass production must not affect protected or vulnerable biodiversity and will, where possible, have to strengthen biodiversity.
Theme 4: environment	Principle 5: In the production and processing of biomass the soil and the soil quality are retained or improved. Principle 6: In the production and processing of biomass ground and surface water must not be depleted and the water quality must be maintained or improved. Principle 7: In the production and processing of biomass the air quality must be maintained or improved.
Theme 5: prosperity	Principle 8: The production of biomass must contribute towards local prosperity
Theme 6: social well-being	Principle 9: The production of biomass must contribute towards the social well-being of the employees and the local population

Source: Project group 'Sustainable production of biomass' (2007), Testing framework for sustainable biomass, p. 10-19.

that is created, on policy, practice and the proportion of budget spent on local supply companies and on the procedures for the appointment of local staff and the share of local senior management.

Unlike the themes of 'competition with food' and 'prosperity', substantive criteria and indicators have been formulated for the theme of social well-being, such as a specification of the ninth principle (see Box 1). One of the criteria is that the production of biomass should have no negative effects on human rights. The minimum requirement is compliance with the Universal Declaration of Human Rights concerning non-discrimination; freedom of trade organisation; child labour; forced and compulsory labour; disciplinary practices, safety practices and the rights of indigenous people. Another criterion is that the use of land must not lead to the violation of official property and use, and customary law without the free and prior consent of the sufficiently informed local population. The minimum requirements are: no land use without the informed consent of original users; land use must be carefully described and officially laid down; official property and use, and customary law of the indigenous population must be recognised and respected.

The pioneering and challenging work of the Dutch did not go unnoticed. 'The Netherlands was the first in the world to establish guidelines to reduce the environmental impact of biofuels production', says environmental watch Mongabay (2007). The OECD included the full list of principles and criteria in a policy paper, commenting that, 'Taken together, the proposed criteria are extremely stringent and would be a challenge to satisfy, even by many producers in OECD countries' (OECD, 2007: 40). In my view the comprehensive set of criteria is indeed innovative and challenging but very much runs the risk of dying a gentle death or being watered down, for three reasons in particular.

3.1 Exclusive multi-stakeholder consultation

Many different stakeholders were involved in the consultation process: staff of ministries, companies, NGOs and universities. However, they had one thing in common. They were all Dutch. However, the scope of the framework is global: 'The framework is applicable to biomass of all origins, so coming from the EU or from outside the EU'. Having said this, the multi-stakeholder consultation was not that inclusive. Transnational Institute and others (2007) reported that one of the expert members of the Cramer working group had admitted that leaving out local stakeholders was a 'weakness' of the initiative. According to this institute, 'nearly all EU-focused initiatives to develop sustainability certification for biomass or agrofuels have failed to include any civil society stakeholders from producing countries in the South, let alone those groups directly affected by monoculture expansion to produce agrofuels' (Transnational Institute *et al.*, 2007: 20). But other stakeholders, for instance the Government of Brazil, were not consulted in the drafting of the framework either. The Dutch framework is meant to be applicable to biomass of all origins but why would Brazil as a major producer of biofuels take this sustainability framework seriously if it has not been consulted at all.

One reason for not consulting Brazilian authorities or those of other major biofuel producing countries (like Argentina, Malaysia, Indonesia and the USA) could be that the criteria are not meant to govern behaviour of other governments or foreign companies but to help the Dutch government to incorporate sustainability criteria into its policy and to stimulate the further international development of a certification system for biomass. In fact, the Cramer commission explicitly states that the report can be used for these two latter purposes. However, the Netherlands is not a major producer or exporter of biofuel feedstock but a major trading hub and importer. This means that the Cramer criteria cannot just be implemented by the Netherlands but have to be negotiated with other countries. One cannot even be sure whether the Dutch authorities or companies will actually use the criteria at all, as the case of the conclusion of the Brazil-Netherlands memorandum of understanding on bioenergy cooperation clearly suggests (see 3.3).

3.2 Discussing new rules but not the rules of the game

The Dutch framework is comprehensive in addressing different dimensions of sustainability of biomass at the micro and macro level. However, not all themes have been elaborated in the same level of detail. For the food versus fuel theme and the theme of prosperity no criteria have yet been developed. This reflects the lack of understanding and consensus among stakeholders on the nature of complex relationships between biofuels, food security and prosperity. As a result, the Dutch framework is a bit lopsided and has open ends. It is not clear how new criteria will evolve from reporting, what agencies can or should collect (reliable and useful) data and how to attribute changes in land use and food prices to biofuel production.

One may argue that the development of a more balanced framework is a matter of time and research. However, this is to ignore a fundamental flaw of the framework: it does not discuss what could and should be the roles and responsibilities of different state and non-state actors in implementing and further developing the criteria. Likewise, it does not discuss the implications of biofuels as emerging global commodities for the development and implementation of sustainability criteria. Whereas the framework is the result of a (Dutch) multi-stakeholder consultation, no ideas have been proposed on behind-the-border and transborder institutional arrangements, regulatory regimes or modes of governance that are necessary to develop and implement the criteria, not to mention complaint mechanisms, dispute settlement and sanctions. The framework is meant to provide criteria that indicate whether biomass has been produced in a responsible manner but does not address responsibilities, obligations, rights. It proposes new rules but does not reflect on new rules of the game.

One could argue that this is a matter of sequencing: developing criteria first and then discussing responsibilities, obligations and sanctions. However, the fact is that discussion on the latter has not been referred to or announced in the Netherlands' Testing Framework for Sustainable Biomass. As a result, it is quite unrealistic to expect that the criteria can be easily implemented. In my view, the design of sustainability criteria only makes sense if they go hand in hand with the design of institutional arrangements, definition of responsibilities, complaint mechanisms, dispute settlement and sanction mechanisms.

3.3 Within the frameworks of WTO and EU law

Shortly after she had submitted the report to the Dutch parliament as chair of the project group, Jacqueline Cramer became Minister of Housing, Spatial Planning and Environment. This clearly provided her with a great opportunity to turn the guidelines and criteria into official policy. However, when Minister Cramer submitted her report to the Dutch parliament, an interdepartmental working group had just finalised an assessment of possibilities for inclusion of sustainability criteria into government policy (Werkgroep Beleidsinpassing, 2007). A few months later this assessment was submitted by Minister Cramer to the Dutch parliament as well.[288] The report states that any policy measures based on sustainability criteria, developed by the working group led by Professor Cramer, had to be taken within the frameworks of WTO and EU law. Measures that unnecessarily restrict trade should be avoided.

For every principle, the risk that either the principle or minimum criteria would not be considered compatible with WTO or EU law has been assessed by a commission of three legal experts (Bronkers *et al.*, 2007). They considered that the risk of minimum criteria to green house gas emissions being rejected would be low. For the four principles on biodiversity and environment, the risk is considered to be moderate to high. Minimum criteria or requirements to prosperity and social well-being were considered incompatible with WTO and EU law. Strikingly, the report made no mention at all about whether the principle on competition with food supply stands a chance against WTO or EU law.

So far the predictions of the legal experts have proven quite right. However, this does not mean that the themes of food, prosperity and social well-being do not appear in policymaking and drafting proposals. The negotiations on the EU directive on renewable energy illustrate this. Whereas all themes contained in the Cramer report and similar ones from other EU countries are being addressed in these negotiations, some are much more prominent on the agenda and more elaborated in proposals than others. The issues of greenhouse gas emissions and to a lesser extent biodiversity and environmental issues are being well addressed in drafts of the EU directive. Cramer claims victory, assuring Dutch parliament that 'our three most important sustainability criteria have been adopted in Brussels'.[289] For the impact on social sustainability, availability of foodstuffs in exporting countries, the ability of people in developing countries to afford these foodstuffs,

[288] Letter to the House of Commons, TK 30305, nr 34.

[289] House of Commons, extraordinary debate on the Memorandum of Understanding between the Netherlands and Brazil on bioenergy cooperation, including biofuels, TK 80, 23 April 2008.

and land use rights, the proposed texts of the EU directive on renewable energy are not only short but also limited to a reporting requirement. In addition, it is suggested that for each country which is a significant source of raw material for biofuel consumed in the EU, a check should be made as to whether the country has ratified the conventions of the International Labour Organization (including those on abolition of forced labour, child labour, and discrimination in respect of employment and occupation).

The Dutch interdepartmental working group that has assessed the possibilities for inclusion of the Cramer criteria into policy, provides two proposals to the Dutch government with regard to the implementation and further development of sustainability criteria. First, the working group strongly recommends that the Dutch government prompt the business sector to adopt 'voluntary agreements'. The working group argues that these agreements could be used to hold companies accountable but does not explain what kind of self-regulation or accountability mechanism could be put in place. The working group only mentions that various companies, including some large energy and oil companies, have indicated that they want to 'actively, quickly and flexibly' develop such agreements. Second, the working group strongly recommends that the Dutch government start a dialogue with producing countries to establish joint agreements on the sustainability of biomass production. A memorandum of understanding can provide the framework for such a dialogue.

An example is the Memorandum of Understanding between the Netherlands and Brazil on bioenergy cooperation, including biofuels, signed in April 2008. What is striking, however, is that this memorandum does not include any reference to the Cramer report or testing framework on sustainable biomass. When the Dutch parliament called for an extraordinary meeting to discuss this, Minister Cramer together with her colleague from the Ministry of Economic Affairs, explained that the Cramer criteria would be the guidelines and point of departure for the Netherlands in discussing and defining sustainable biomass together with Brazil. When the question was raised as to what sanctions would be taken if the implementation of the joint agreement did not meet one of the Cramer criteria, the Minister of Economic Affairs explained that, 'with respect to another country, we obviously can not impose sanctions. However, as part of the discussion on the implementation of the Memorandum of Understanding, some things can be done, others not'.[290]

The Netherlands has been a front-runner in developing a global framework with guidelines and criteria for sustainable biomass of all origins, but considers them negotiable when concluding new agreements with emerging agricultural powers and energy exporters. The right to food and human rights in general are not the

[290] *ibid.* TK 80, 23 April 2008, pg. 5677, my translation.

point of departure or normative framework for Dutch international cooperation in the field of biofuels. This is very much in line with the observations of Oshaug, Van der Meulen and Vlemminx (see this volume) who each show in a different way that the Netherlands is fed up with the right to food in the making of law at the domestic and international level.

4. Conclusion

It makes a lot of difference whether sustainability or the right to food is taken as point of departure when developing biofuel policies and regulating biofuel production. Framing and judging biofuel policy in terms of sustainability, as proposed by the UN-energy group and the Dutch Cramer Commission, is comprehensive in that it addresses the environmental, social as well as economic impact of biofuel production. It may serve as a warning to governments not to see biofuels merely as an opportunity for economic development or an alternative source of energy but also as a potential threat to the environment, biodiversity, human rights and access to food.

However, there is a risk that the three social sustainability guidelines and criteria of the Cramer Commission in particular (on access to food, prosperity and social well-being) may die a gentle death or be watered down if and when the possibility to include them in government policy is only tested in terms of their compatibility with EU and WTO law and particularly in terms of the economic principle of unnecessary restriction of trade. It is quite remarkable that the proposed sustainability criteria of the Cramer Commission were assessed in terms of their compatibility with EU and WTO law but not in terms of international human rights law.

To lower the risk of social sustainability considerations being squeezed in negotiations on sustainable biofuels or renewable energy, it is important that development NGOs and human rights defenders (like Eide) continue to remind government officials of their broad sustainability agenda at the national and international level. Still, this is to uphold the notion of sustainability and its dimensions (including respect for human rights) as political commodities. One dimension can be traded for another in EU negotiations, in private regulation or voluntary agreements of a commercial sector, or bilateral agreements. Putting principles and procedures of the human right to food upfront in negotiations and policymaking on sustainable biomass, as proposed by Eide, will not make negotiations easier and predictable. However, this can help prevent the three social sustainability criteria from falling off the agenda and national governments, like the Netherlands, from hiding behind a very shallow sustainability discourse when concluding bilateral agreements and promoting voluntary agreements by the commercial sector.

If we undo the body of international law on the right to food from its normative and prescriptive dimensions and focus on its procedural ones, a three-fold research and academic agenda evolves. The first issue is to study the effects of biofuel production on different, more or less vulnerable segments of society. The second is to explore how different ministries of a country define their mandate, responsibility and obligations with regard to biofuels and why the right to food is or is not used for this purpose. The third is about the implications of the emergence of biofuels as a global commodity for human rights doctrine and transborder obligations of states.

References

Bronkers, M., Verberne, G. and Juttmann, Ph. 2007. WTO/EG-rechtelijke toetsing van de door de projectgroep duurzame productie van biomassa opgestelde duurzaamheidscriteria. Annex of letter to the Dutch House of Commons (TK 30305, nr. 34).

D'Souza, F. 1994. Democracy as a cure for famine. *Journal of Peace Research* 31(4): 369-373.

Eide, A. 2008. The right to food and the impact of liquid biofuels (agrofuels). FAO Right to Food Studies.

European Parliament and European Council. 2003. Directive 2003/30/EC on the promotion of the use of biofuels or other renewable fuels for transport

EU Proposal for a Directive of the European Parliament and of the Council amending Directive 98/70/EC as regards the specification of petrol, diesel and gas-oil and introducing a mechanism to monitor and reduce greenhouse gas emissions from the use of road transport fuels and amending Council Directive 1999/32/EC, as regards the specification of fuel used by inland waterway vessels and repealing Directive 93/12/EEC, 27 June 2008.

EU Proposal for a Directive of the European Parliament and of the Council on the promotion of the use of energy from renewable sources, 27 June 2008.

EU Presidency suggestions for a common sustainability scheme, 27 June 2008.

Government of the Netherlands and Government of the Federative Republic of Brazil, Memorandum of Understanding on Bioenergy Cooperation, including Biofuels. 11 April 2008.

Hospes, O. 2008a. Overcoming barriers to the implementation of the Right to Food. *European Food and Feed Law Review,* 3(4): 246-261.

Hospes, O. 2008b. Biofuels and food security. Strategy and Policy Brief no. 10. Markets, trade and sustainable rural development. Wageningen International.

Mongabay, 2007. Dutch plan restricts biofuels that damage environment. Http://news.mongabay.com/2007/0429-dutch.html, 29 April 2007.

OECD, 2007. Is the cure worse than the disease? Report of the Round Table on Sustainable Development of the OECD.

OECD and FAO, 2007. OECD-FAO Agricultural Outlook 2007-2016.

Project group Sustainable production of biomass, 2007. Testing framework for sustainable biomass. Report by order of the Netherlands Government.

Sen, A. 1981. Poverty and Famines: An Essay on Entitlement and Deprivation. Oxford, Clarendon Press.

Senauer, B. 2008. The appetite for biofuels starves the poor. www.guardian.co.uk/commentisfree/2008/jul/03/biofuels.usa

House of Commons, Extra-parliamentary debate on the Memorandum of Understanding between the Netherlands and Brazil on bioenergy cooperation, including biofuels. TK 80, pg. 5668-5683, 23 April 2008.

Transnational Institute, Corporate Europe Observatory and Grupo de Reflexión Rural, 2007. Paving the way for agrofuels: EU policy, sustainability criteria and climate calculations.

UN-Energy Group, 2007. Sustainable bioenergy: a framework for decision makers.

Von Braun, J. von. 2007. The world food situation: new driving forces and required actions. Food Policy Report no. 18.

Werkgroep Beleidsinpassing in overleg met IPE/VROM/EZ, 2007. Beleidsinpassing van duurzaamheidscriteria voor biomassa voor biobrandstoffen en elektriciteitsopwekking.

World Bank, 2007. Rising food prices: policy options and World Bank Response.

Chapter 9
The hungry challenging the global elite
Extraterritorial state obligations under the Human Right to Food

Ute Hausmann

1. Introduction

The majority of the 923 million hungry people are living in the countryside, dependent on agriculture, either as small-scale farmers or as agricultural labourers. The agricultural sector and especially small-scale farmers have been hard hit by structural adjustment policies in the 1980s and 1990s, designed and implemented by the International Monetary Fund (IMF) and the World Bank. They have lost state support as well as market access. At the same time, negotiations within the World Trade Organisation (WTO) have led to a far-reaching liberalisation of international trade in agricultural products. The livelihood of millions of peasants worldwide is threatened by cheap agricultural imports because they lose their markets and income. Currently, the European Union (EU) is negotiating free trade agreements in the form of Economic Partnership Agreements (EPA) with the ACP countries, which will further increase access to markets in the global South for European products. The United States of America (USA) and China are also very active in creating overseas markets for their agricultural products ranging from rice to milk and tomato paste. But it is not only unfair competition over markets that is cementing and creating hunger in the rural parts of this world. Increasing competition over land, water and forests is another worrying development which is likely to increase hunger and malnutrition in the coming years. This competition is substantially fuelled by an increasing international demand for agricultural and horticulture products, for timber, oil and minerals, and lately also for agrofuel. Access to and the licence to exploit the land is essential for international investors, which have the backing of their home governments and who are in part very powerful in negotiating investment agreements with the host government, which can be enforced though international dispute settlement.

Freedom from hunger is one of the fundamental human rights. The human right to food has been recognised in the Universal Declaration of Human Rights as well as several human rights treaties, the most important of which is the International Covenant on Economic, Social and Cultural Rights (ICESCR). International human rights treaties carry the spirit that the full realisation of human rights is possible and that realising human rights is a common purpose of the international community. But for a very long time, hunger as an international concern has been discussed almost exclusively as a matter of supporting national efforts to combat hunger through development cooperation or food aid. Seldom has the international community criticised a national government for violating the human right to food and for not doing enough to combat hunger in its country. However, what is most astonishing is that the international community has so far failed to take a close look at the causes of hunger and violations of the human right to food that lie outside a state's boundaries. One of today's key challenges for human rights academics and activists lies in strengthening both the interpretation and implementation of states' extraterritorial human rights obligations and ultimately making these obligations legally enforceable, presenting victims of human rights violations with an effective remedy.

2. An international economic order based on the human right to food

Following the *World Food Summit plus five* in 2002, the member states of the Food and Agriculture Organisation of the United Nations (FAO) started negotiations on *Voluntary Guidelines to support the progressive realisation of the right to adequate food in the context of national food security*. As is implied by the full title of these Voluntary Guidelines, the 'donor countries' in particular were eager to avoid discussions on aspects like international trade, food aid or the level of development assistance, and to restrict the guidelines to address only national policies. However, developing countries, especially the African states, entered the negotiations with the very clear demand that these issues be addressed. As a compromise, a part was added to the Voluntary Guidelines that addressed the international dimension of the human right to food. This part is basically a repetition of earlier international agreements on trade and aid. Still, the Voluntary Guidelines, which were adopted unanimously by the FAO Council in November 2004, are a very significant development in the international human rights debate. Firstly, this was the first time that governments negotiated about the policy implications of the human right to food as it is protected in the Universal Declaration of Human Rights and the International Covenant on Economic, Social and Cultural Rights (ICESCR). The negotiators also confirmed the legal interpretation of the human right to food in the way it has been developed by the UN Committee on economic, social and cultural Rights (CESCR) in its General Comment No. 12 of 1999. Secondly, the Voluntary Guidelines are significant because they recognise in principle the international forces that can support or undermine national efforts to implement the human right to food. However, this recognition is very vague and not very helpful as it takes international agreements on trade and aid as a starting point rather than the ICESCR. In this way it is almost impossible to challenge existing trade and investment regimes that are a major factor in keeping the world hungry. The Voluntary Guidelines are only one example of the disconnection between the general recognition of the supremacy of human rights in international law on the one hand and the lack of teeth of international human rights instruments regarding their ability to shape the international economic order. The international community has so far failed to implement its commitment made in Article 28 of the Universal Declaration of Human Rights: 'Everyone is entitled to a social and international order in which the rights and freedoms set forth in this Declaration can be fully realized.'

International human rights law is often understood to address only the relationship between the state that has ratified human rights treaties and the individuals and groups who live in this state. In the same line, human rights treaties are understood as guiding national policies only when they address internal policies. When taking a look at the International Covenant on Economic, Social and Cultural Rights (ICESCR), it becomes obvious that this is a clear misreading of existing human rights law. The intention of the ICESCR was to cast in international binding law human rights obligations of individual states in relation to the universal implementation of Economic, Social and Cultural rights, based on commitments made by the international community in the Universal Declaration of Human Rights. In addition, the preamble of the ICESCR states that this treaty is part of the effort to implement 'the obligation of states under the Charter of the United Nations to promote universal respect for, and observance of human rights and freedoms'. The ICESCR emphasises repeatedly the importance of international cooperation to achieve the full realisation of the rights recognised in the treaty, including the human right to food. It is therefore obvious that states which ratify the ICESCR have the obligation to cooperate internationally to achieve the full realisation of the human right to food worldwide. In its legal interpretation of the human right to food in General Comment No. 12, the UN Committee on Economic, Social and Cultural Rights (CESCR) has emphasised that states parties to the ICESCR should 'respect the enjoyment of the right to food in other countries, to protect that right, to facilitate access to food and to provide the necessary aid when required'. In his report to the UN Commission for Human Rights in 2005 the UN Special Rapporteur on the Right to Food has further elaborated on the extraterritorial obligations of states under the ICESCR:

- *The obligation to respect* requires states to ensure that their own policies and practices do not lead to the violation of the right to food in other countries.
- *The obligation to protect* requires states to ensure that their own citizens and companies, as well as other third parties subject to their jurisdiction, including transnational corporations, do not violate the right to food in other countries.
- *The obligation to support the fulfillment* of the right to food requires states, depending on the availability of resources, to facilitate the realisation of the right to food in other countries and to provide the necessary aid when required.

Much needs to be done to concretise and operationalise these obligations. Governments are very reluctant to tackle the question of what extraterritorial obligations under the human right to food imply for their own policies. Some experience from Germany can serve to highlight some of the crucial issues.

3. Human Rights always start at home

Compliance with extraterritorial obligations under the human right to food requires initiatives cutting across various policy fields and institutions. Implementation of these obligations requires unilateral action as well as bi- and multilateral cooperation. These three levels of implementation are closely interwoven. Unilateral action is most obviously required regarding the obligation to respect the human right to food. This means that the state has to ensure that policies that are designed in a national context will not have a negative impact on the human right to food. Due diligence standards in relation to human rights prescribe procedures which are intended to prevent human rights violations by the state itself, by private actors or by international organisations of which the state is a member. The state has to ensure that it does not – by act or omission – cause violations of the human right to food in other countries and that it does not tolerate or facilitate human rights abuses by other actors. A failure to exercise such due diligence and to act accordingly constitutes a violation of its obligations under the ICESCR. However, legislative procedures are not usually designed to address the possible impact on the human right to food in other countries. In its 5th state report on the implementation of the ICESCR, the German government states that in general it is possible to address the impact of new legislation on international economic, social and cultural human rights as part of the Regulatory Impact Assessment (RIA). However, there is no evidence that this is also being practiced. It is therefore promising that the government is currently revising its handbook for lawmakers on how to ensure compliance of legislation with higher ranking law, including international human rights treaties. In addition, the government has announced in its human rights action plan for the years 2008 to 2010, that it will 'increasingly assume responsibility for the right to food. This is also related to regulation on the national and international level which has implications for global food security.'

4. Human Rights – green light for export and investment promotion?

A political playing field that is particularly at risk of contributing to violations of the right to food in other countries is that of export and investment promotion. Generally there are two aspects that need to be addressed: first, it is obvious that a state should not actively support activities of companies which are violating human rights in other countries, for example by providing risk insurance, financial support or political backing by embassy personnel in the respective country. A recent example in which governments had to demonstrate their commitment to human rights is the granting of export guarantees by the German, Austrian and Swiss Export Credit Agencies (ECAs) to companies from these countries which are involved in constructing the Ilisu dam in Turkey. The Ilisu Dam on the Tigris will displace 55,000-65,000 Kurdish people, drown the 10,000-year-old city of Hasankeyf, and cause extensive environmental damage, all of which presents a substantial threat to the right to food of the affected population. The project faces strong opposition from the local population and civil society groups. Still, the ECAs went ahead and approved the guarantees for the project with 153 conditions attached. When in March 2008 and then again in autumn 2008 an independent monitoring team presented a devastating report on the resettlement process which indicates that the right to food has been violated substantially, civil society organisations called on the governments of Germany, Austria and Switzerland to withdraw the guarantees. Withdrawal of the guarantees would most likely end the involvement of the German, Austrian and Swiss companies in the project and would send a strong message to the Turkish government as well as potential investors. In early October 2008 the three governments announced that they had given the Turkish government sixty days to comply with the 153 conditions before the guarantee would be cancelled.

The second aspect is that it might not be the investment itself that is problematic in human rights terms but rather the regulatory framework that is created to promote foreign direct investment. Bilateral investment agreements between states are increasingly being scrutinised by human rights activists and lawyers. In 2006, the investment agreement between Germany and Paraguay even became a matter for discussion before the Inter-American Court for Human Rights. One of the key purposes of investment agreements is to protect foreign investments from arbitrary expropriation by the host state. However, in many developing countries like Paraguay the constitution provides for expropriation in the public interest. Expropriation is in many states a necessary instrument to redistribute idle land to landless peasants and to thereby fulfil the right to food. In Paraguay, one percent of the population owns seventy-seven percent of the land. This is the highest concentration of land in the hands of few in Latin America. The unequal distribution of land is a major reason for the fact that fourteen percent of the population are suffering from hunger. A substantial acreage of land in Paraguay is owned by German citizens, much of it was bought for speculative reasons by German citizens who do not even live in Paraguay. Even though this land qualifies for redistribution – against compensation – under the national law on agrarian reform, no expropriations have taken place since the investment agreement between Paraguay and Germany entered into force in 1993. Apparently the authorities in Paraguay are afraid that German land owners would drag the state before an international dispute settlement body and that the state would have to pay huge penalties. In 2006, the Inter-American Court for Human Rights heard the case of the indigenous Sawhoyamaxa. Since 1991 they had been struggling to get back 14,000 hectare of their traditional land which was now in the hands of the German citizen Heribert Roedel. Even though the state institute for agrarian reform had recommended the transfer of the land to the indigenous group, the senate refused to sign the corresponding legislation based on the argument that this was not possible under the investment agreement with Germany. In its assessment of the case, the Inter-American Court for Human Rights came to the conclusion that the state had to transfer the land in order to comply with the constitution as well as human rights. It also stated that bilateral investment agreements cannot overrule either constitutional provisions or international human rights treaties. The state of Paraguay was ordered to redistribute the land within two years and to pay compensation for the loss of life of eighteen children who died of hunger and malnutrition as result of the landlessness of the families. Today, the two years have almost passed but the families are still not living on the land. In order to address this serious negative impact of the bilateral investment agreement on land reform and the right to food in Paraguay, FIAN as well as the German Protestant and Catholic development organisations Misereor and Brot für die Welt are calling on the government of Germany to send an official note to the government of Paraguay clarifying that Germany respects the expropriation of land in order to fulfil the right to food in Paraguay. So far, the German government has refused to do so.

5. Home state regulation of transnational economic activities

One issue that is increasingly being discussed is the extent to which a state is obliged to ensure that private companies headquartered in this state do not violate human rights or become complicit in human rights violations in other countries. Two aspects are relevant here: first, states might be obliged to enact regulation which will prevent the violation of the right to food by private economic interest. A very topical example is the control of speculation in grains which is partly responsible for the food price hike in 2007/2008. The second related aspect is the question as to whether the home state is obliged to ensure that victims of human rights violations in connection with activities of transnational companies (TNCs) have access to effective remedies. The coffee plantation in Uganda, operated by a subsidiary of the German Neumann Kaffee Gruppe, can serve as an example. In August 2001, the government of Uganda forcefully evicted 2,000 farmers from their land in the Mubende district to make way for a coffee plantation operated by the Kaweri Coffee Plantation Ltd., owned by Neumann Kaffee Gruppe. Only a few days after the eviction, the plantation was opened in a ceremony with the participation of the head of the Neumann Kaffee Gruppe. To date, the farmers have neither been fully compensated nor do they have access to new farm land. The farmers finally decided to go to court against the government of Uganda as well as the Kaweri Coffee Plantation Ltd. However, there are many obstacles. It took a long time and international pressure to get the court process started. Also, the speaker of the community, a teacher, faced massive intimidation and was even jailed for several months on false charges. The community members are prepared to fight for justice, but there is no guarantee that the courts will award them justice. If the victims of human rights violations do not get justice, shouldn't it be possible to go to court against a TNC in its home country? Many lawyers and civil society activist believe that this could prove an effective tool to remedy and eventually prevent human rights violations. The first step to implement this would be to create the necessary legal basis to try TNCs for human rights violations in other countries.

6. Liberalisation versus Human Rights

International trade in agricultural products has become a threat to the human right to food in many countries. Eighty percent of those hungry today are living in the countryside, most of them small-scale farmers, many others landless agricultural labourers. The structural adjustment policies (SAP) imposed by the World Bank and the IMF since the 1980s have had a devastating effect on the ability of peasants to earn a decent living from agriculture. The SAPs implied that most of the state support to agriculture was eliminated. At the same time, the US and Europe were dumping agricultural products (often disguised as food aid) on the markets of developing countries at prices which were so low that no local farmer could compete. While more attention has been given in the last years to reduce export subsidies to producers in the US and Europe, farmers in these countries still receive very strong state support, something that farmers in developing countries can only dream of. And, even though governments and the Bretton Woods institutions have acknowledged the devastating effects of the SAPs, trade liberalisation is still at the top of the agenda. A particular challenge that African countries are facing at the moment are the negotiations on Economic Partnership Agreements (EPA) with the European Union (EU). Investigations by FIAN and others have shown that the further opening of African markets to agricultural imports from the EU would dramatically impact on the income and thereby on the right to food of tomato farmers in Ghana or milk producers in Zambia. The governments of these states would lose the ability to protect small-scale farmers from cheap imports. In addition, the recent food price crisis has demonstrated that the ever-increasing dependence of developing countries on agricultural imports can have substantial effects on the right to food in these countries which are almost impossible to compensate. In this context what are the extraterritorial human rights obligations of the EU member states? The UN Committee on Economic, Social and Cultural Rights (CESCR) has taken the approach that when governments negotiate international agreements, they have to acknowledge the supremacy of human rights law and to ensure that these agreements are in line with the states' obligations.

While international agreements represent legal documents that may in the future be challenged in national courts on human rights grounds, it is more difficult to assess the impact on the right to food of policy advice given by the Bretton Woods institutions. The World Bank and the IMF are involved in almost all developing countries, advising governments on how to make their country more attractive to foreign direct investment or how to integrate into the world economy. Resulting legislative changes often fail to respect the right to food. An example is the pressure exercised by the IMF on the government of Ghana when the government decided to increase import taxes on selected agricultural products so that local producers could stay competitive. In this case, the IMF admitted in writing that it had advised the government on the issue. In most other cases, the pressure exercised and the tactics employed by the Bretton Woods institutions are more disguised. In 2005, the UN General Assembly called on 'all relevant international organisations, including the World Bank and the International Monetary Fund, to promote policies and projects that have a positive impact on the right to food, to ensure that partners respect the right to food in the implementation of common projects, to support strategies of member states aimed at the fulfilment of the human right to food and to avoid any actions that could have a negative impact on the realisation of the right to food'. Also, for some years now the CESCR has asked governments as members of international financial institutions, in particular the International Monetary Fund and the World Bank, to do all they can to ensure that the policies and decisions of those organisations are in conformity with the obligations of States parties to the International Covenant on Economic, Social and Cultural Rights. In relation to the Bretton Woods institutions, the human rights movement faces two challenges: first, how to ensure that the states which rule these institutions take decisions based on human rights and enforce human rights standards in internal decision-making processes. Second, the international community will ultimately have to deal with the question of how to initiate external accountability mechanisms which will enable victims of human rights violations to challenge these powerful institutions if they are responsible for their suffering.

7. Working together to realise the human right to food for all

While most of the aspects discussed above deal with the extraterritorial obligation not to do harm, it is important not to forget that the International Covenant on Economic, Social and Cultural Rights obliges states parties to actively cooperate to realise the rights protected in the Covenant. One of the most beautifully phrased recognitions of this obligation can be found in the Human Rights Action Plan of the German Ministry for Economic Cooperation and Development for the years 2008 to 2010: 'Human rights provide us with legally binding standards to which we, in common with our partner countries, have committed ourselves inside and outside our borders. We have jointly ratified international human rights treaties and so it is our joint responsibility to work for the respect, protection and fulfilment of human rights. By meeting our obligations, we want to help our partners specifically and effectively to meet theirs.' While this may sound visionary to many, the vision would not be complete without envisaging that one day victims of violations of the human right to food will be able to challenge before a court not only their own governments, but also those of foreign states, companies headquartered in these states and international financial institutions ruled by these states. Those who today are being criminalised for being poor and hungry will stand up and start to address foreign states and companies for obstructing the realisation of the human right to food for all.

Further reading

Brot für die Welt, eed, FIAN 'Germany's extraterritorial Human Rights Obligations. Introduction and six case studies', 2006 www.fian.org/resources/documents/others/germanys-extraterritorial-human-rights-obligations/pdf.

Brot für die Welt, eed, FIAN 'Germany's extraterritorial human rights obligations in multilateral development banks. Introduction and case study of three projects in Chad, Ghana and Pakistan', 2006, www.fian.org/resources/documents/others/germanys-extraterritorial-human-rights-obligations-in-multilateral-development-banks/pdf.

Brot für die Welt, eed, FIAN 'Wie deutscher Landbesitz in Paraguay effektive Hungerbekämpfung verhindert', 2007, www.fian.de/fian/downloads/pdf/agrar/deutscher_Landbesitz_Paraguay_07_deu.pdf.

Ecumenical Advocacy Alliance 'Trade Policies and Hunger: The Impact of Trade Liberalisation on the Right to Food of Rice Farming Communities in Ghana, Honduras and Indonesia', 2007, www.e-alliance.ch/globaltrade/policiesandhunger.pdf.

FIAN 'Right to Food of Tomato and Poultry Farmers. Report of an investigative mission to Ghana', 2007, www.fian.de/fian/downloads/pdf/agrarhandel/report_ghana.pdf.

FIAN 'Time for a Human Rights Framework for Action. Position Paper on the Comprehensive Framework for Action of the High Level Task Force on the Food Crisis', 2008. www.fian.org/resources/documents/others/time-for-a-human-right-to-food-framework-of-action/pdf.

Annexes

Annex 1 Treaty provisions on the human right to food

Universal Declaration of Human rights (UDHR)

Article 25

1. Everyone has the right to a standard of living adequate for the health and well-being of himself and of his family, including food, clothing, housing and medical care and necessary social services, and the right to security in the event of unemployment, sickness, disability, widowhood, old age or other lack of livelihood in circumstances beyond his control.
2. Motherhood and childhood are entitled to special care and assistance. All children, whether born in or out of wedlock, shall enjoy the same social protection.

Article 29

1. Everyone has duties to the community in which alone the free and full development of his personality is possible.
2. In the exercise of his rights and freedoms, everyone shall be subject only to such limitations as are determined by law solely for the purpose of securing due recognition and respect for the rights and freedoms of others and of meeting the just requirements of morality, public order and the general welfare in a democratic society.
3. These rights and freedoms may in no case be exercised contrary to the purposes and principles of the United Nations.

International Covenant on Economic, Social and Cultural Rights (ICESCR)

Article 2

1. Each State Party to the present Covenant undertakes to take steps, individually and through international assistance and co-operation, especially economic and technical, to the maximum of its available resources, with a view to achieving progressively the full realization of the rights recognized in the present Covenant by all appropriate means, including particularly the adoption of legislative measures.
2. The States Parties to the present Covenant undertake to guarantee that the rights enunciated in the present Covenant will be exercised without discrimination of any kind as to race, colour, sex, language, religion, political or other opinion, national or social origin, property, birth or other status.
3. Developing countries, with due regard to human rights and their national economy, may determine to what extent they would guarantee the economic rights recognized in the present Covenant to non-nationals.

Article 11

1. The States Parties to the present Covenant recognize the right of everyone to an adequate standard of living for himself and his family, including adequate food, clothing and housing, and to the continuous improvement of living conditions. The States Parties will take appropriate steps to ensure the realization of this right, recognizing to this effect the essential importance of international co-operation based on free consent.
2. The States Parties to the present Covenant, recognizing the fundamental right of everyone to be free from hunger, shall take, individually and through international co-operation, the measures, including specific programmes, which are needed:
   a. To improve methods of production, conservation and distribution of food by making full use of technical and scientific knowledge, by disseminating knowledge of the principles of nutrition and by developing or reforming agrarian systems in such a way as to achieve the most efficient development and utilization of natural resources;
   b. Taking into account the problems of both food-importing and food-exporting countries, to ensure an equitable distribution of world food supplies in relation to need.

(International) Convention on the Rights of the Child (ICRC)

Article 27

1. States Parties recognize the right of every child to a standard of living adequate for the child's physical, mental, spiritual, moral and social development.
2. The parent(s) or others responsible for the child have the primary responsibility to secure, within their abilities and financial capacities, the conditions of living necessary for the child's development.
3. States Parties, in accordance with national conditions and within their means, shall take appropriate measures to assist parents and others responsible for the child to implement this right and shall in case of need provide material assistance and support programmes, particularly with regard to nutrition, clothing and housing.
4. States Parties shall take all appropriate measures to secure the recovery of maintenance for the child from the parents or other persons having financial responsibility for the child, both within the State Party and from abroad. In particular, where the person having financial responsibility for the child lives in a State different from that of the child, States Parties shall promote the accession to international agreements or the conclusion of such agreements, as well as the making of other appropriate arrangements.

Annex 2 General Comments of UN Committee on Economic, Social and Cultural Rights

General Comment 3

The nature of States parties obligations
(Art. 2, para. 1 of the Covenant)
(E/1991/23)

1. Article 2 is of particular importance to a full understanding of the Covenant and must be seen as having a dynamic relationship with all of the other provisions of the Covenant. It describes the nature of the general legal obligations undertaken by States parties to the Covenant. Those obligations include both what may be termed (following the work of the International Law Commission) obligations of conduct and obligations of result. While great emphasis has sometimes been placed on the difference between the formulations used in this provision and that contained in the equivalent Article 2 of the International Covenant on Civil and Political Rights, it is not always recognized that there are also significant similarities. In particular, while the Covenant provides for progressive realization and acknowledges the constraints due to the limits of available resources, it also imposes various obligations which are of immediate effect. Of these, two are of particular importance in understanding the precise nature of States parties obligations. One of these, which is dealt with in a separate general comment, and which is to be considered by the Committee at its sixth session, is the 'undertaking to guarantee' that relevant rights 'will be exercised without discrimination...'.
2. The other is the undertaking in Article 2 (1) 'to take steps', which in itself, is not qualified or limited by other considerations. The full meaning of the phrase can also be gauged by noting some of the different language versions. In English the undertaking is 'to take steps', in French it is 'to act' ('s'engage à agir') and in Spanish it is 'to adopt measures' ('a adoptar medidas'). Thus while the full realization of the relevant rights may be achieved progressively, steps towards that goal must be taken within a reasonably short time after the Covenant's entry into force for the States concerned. Such steps should be deliberate, concrete and targeted as clearly as possible towards meeting the obligations recognized in the Covenant.
3. The means which should be used in order to satisfy the obligation to take steps are stated in Article 2 (1) to be 'all appropriate means, including particularly the adoption of legislative measures'. The Committee recognizes that in many instances legislation is highly desirable and in some cases may even be indispensable. For example, it may be difficult to combat discrimination effectively in the absence of a sound legislative foundation for the necessary measures. In fields such as health, the protection of children and mothers, and education, as well as in respect of the matters dealt with in Articles 6 to 9, legislation may also be an indispensable element for many purposes.
4. The Committee notes that States parties have generally been conscientious in detailing at least some of the legislative measures that they have taken in this regard. It wishes to emphasize, however, that the adoption of legislative measures, as specifically foreseen

by the Covenant, is by no means exhaustive of the obligations of States parties. Rather, the phrase 'by all appropriate means' must be given its full and natural meaning. While each State party must decide for itself which means are the most appropriate under the circumstances with respect to each of the rights, the 'appropriateness' of the means chosen will not always be self-evident. It is therefore desirable that States parties' reports should indicate not only the measures that have been taken but also the basis on which they are considered to be the most 'appropriate' under the circumstances. However, the ultimate determination as to whether all appropriate measures have been taken remains one for the Committee to make.

5. Among the measures which might be considered appropriate, in addition to legislation, is the provision of judicial remedies with respect to rights which may, in accordance with the national legal system, be considered justiciable. The Committee notes, for example, that the enjoyment of the rights recognized, without discrimination, will often be appropriately promoted, in part, through the provision of judicial or other effective remedies. Indeed, those States parties which are also parties to the International Covenant on Civil and Political Rights are already obligated (by virtue of arts. 2 (paras. 1 and 3), 3 and 26) of that Covenant to ensure that any person whose rights or freedoms (including the right to equality and non-discrimination) recognized in that Covenant are violated, 'shall have an effective remedy' (Art. 2 (3) (a)). In addition, there are a number of other provisions in the International Covenant on Economic, Social and Cultural Rights, including Articles 3, 7 (a) (i), 8, 10 (3), 13 (2) (a), (3) and (4) and 15 (3) which would seem to be capable of immediate application by judicial and other organs in many national legal systems. Any suggestion that the provisions indicated are inherently non-self-executing would seem to be difficult to sustain.
6. Where specific policies aimed directly at the realization of the rights recognized in the Covenant have been adopted in legislative form, the Committee would wish to be informed, inter alia, as to whether such laws create any right of action on behalf of individuals or groups who feel that their rights are not being fully realized. In cases where constitutional recognition has been accorded to specific economic, social and cultural rights, or where the provisions of the Covenant have been incorporated directly into national law, the Committee would wish to receive information as to the extent to which these rights are considered to be justiciable (i.e. able to be invoked before the courts). The Committee would also wish to receive specific information as to any instances in which existing constitutional provisions relating to economic, social and cultural rights have been weakened or significantly changed.
7. Other measures which may also be considered 'appropriate' for the purposes of Article 2 (1) include, but are not limited to, administrative, financial, educational and social measures.
8. The Committee notes that the undertaking 'to take steps ... by all appropriate means including particularly the adoption of legislative measures' neither requires nor precludes any particular form of government or economic system being used as the vehicle for the steps in question, provided only that it is democratic and that all human rights are thereby respected. Thus, in terms of political and economic

systems the Covenant is neutral and its principles cannot accurately be described as being predicated exclusively upon the need for, or the desirability of a socialist or a capitalist system, or a mixed, centrally planned, or laisser-faire economy, or upon any other particular approach. In this regard, the Committee reaffirms that the rights recognized in the Covenant are susceptible of realization within the context of a wide variety of economic and political systems, provided only that the interdependence and indivisibility of the two sets of human rights, as affirmed inter alia in the preamble to the Covenant, is recognized and reflected in the system in question. The Committee also notes the relevance in this regard of other human rights and in particular the right to development.

9. The principal obligation of result reflected in Article 2 (1) is to take steps 'with a view to achieving progressively the full realization of the rights recognized' in the Covenant. The term 'progressive realization' is often used to describe the intent of this phrase. The concept of progressive realization constitutes a recognition of the fact that full realization of all economic, social and cultural rights will generally not be able to be achieved in a short period of time. In this sense the obligation differs significantly from that contained in Article 2 of the International Covenant on Civil and Political Rights which embodies an immediate obligation to respect and ensure all of the relevant rights. Nevertheless, the fact that realization over time, or in other words progressively, is foreseen under the Covenant should not be misinterpreted as depriving the obligation of all meaningful content. It is on the one hand a necessary flexibility device, reflecting the realities of the real world and the difficulties involved for any country in ensuring full realization of economic, social and cultural rights. On the other hand, the phrase must be read in the light of the overall objective, indeed the raison d'être, of the Covenant which is to establish clear obligations for States parties in respect of the full realization of the rights in question. It thus imposes an obligation to move as expeditiously and effectively as possible towards that goal. Moreover, any deliberately retrogressive measures in that regard would require the most careful consideration and would need to be fully justified by reference to the totality of the rights provided for in the Covenant and in the context of the full use of the maximum available resources.

10. On the basis of the extensive experience gained by the Committee, as well as by the body that preceded it, over a period of more than a decade of examining States parties' reports the Committee is of the view that a minimum core obligation to ensure the satisfaction of, at the very least, minimum essential levels of each of the rights is incumbent upon every State party. Thus, for example, a State party in which any significant number of individuals is deprived of essential foodstuffs, of essential primary health care, of basic shelter and housing, or of the most basic forms of education is, prima facie, failing to discharge its obligations under the Covenant. If the Covenant were to be read in such a way as not to establish such a minimum core obligation, it would be largely deprived of its raison d'être. By the same token, it must be noted that any assessment as to whether a State has discharged its minimum core obligation must also take account of resource constraints applying within the country concerned. Article 2 (1) obligates each State party to take the necessary

steps 'to the maximum of its available resources'. In order for a State party to be able to attribute its failure to meet at least its minimum core obligations to a lack of available resources it must demonstrate that every effort has been made to use all resources that are at its disposition in an effort to satisfy, as a matter of priority, those minimum obligations.

11. The Committee wishes to emphasize, however, that even where the available resources are demonstrably inadequate, the obligation remains for a State party to strive to ensure the widest possible enjoyment of the relevant rights under the prevailing circumstances. Moreover, the obligations to monitor the extent of the realization, or more especially of the non-realization, of economic, social and cultural rights, and to devise strategies and programmes for their promotion, are not in any way eliminated as a result of resource constraints. The Committee has already dealt with these issues in its General Comment 1 (1989).
12. Similarly, the Committee underlines the fact that even in times of severe resources constraints whether caused by a process of adjustment, of economic recession, or by other factors the vulnerable members of society can and indeed must be protected by the adoption of relatively low-cost targeted programmes. In support of this approach the Committee takes note of the analysis prepared by UNICEF entitled 'Adjustment with a human face: protecting the vulnerable and promoting growth[291], the analysis by UNDP in its Human Development Report 1990[292] and the analysis by the World Bank in the World Development Report 1990[293].
13. A final element of Article 2 (1), to which attention must be drawn, is that the undertaking given by all States parties is 'to take steps, individually and through international assistance and cooperation, especially economic and technical...'. The Committee notes that the phrase 'to the maximum of its available resources' was intended by the drafters of the Covenant to refer to both the resources existing within a State and those available from the international community through international cooperation and assistance. Moreover, the essential role of such cooperation in facilitating the full realization of the relevant rights is further underlined by the specific provisions contained in Articles 11, 15, 22 and 23. With respect to Article 22 the Committee has already drawn attention, in General Comment 2 (1990), to some of the opportunities and responsibilities that exist in relation to international cooperation. Article 23 also specifically identifies 'the furnishing of technical assistance' as well as other activities, as being among the means of 'international action for the achievement of the rights recognized...'.
14. The Committee wishes to emphasize that in accordance with Articles 55 and 56 of the Charter of the United Nations, with well-established principles of international law, and with the provisions of the Covenant itself, international cooperation for development and thus for the realization of economic, social and cultural rights is an obligation of all States. It is particularly incumbent upon those States which are

[291] G.A. Cornia, R. Jolly and F. Stewart, eds., Oxford, Clarendon Press, 1987.
[292] Oxford, Oxford University Press, 1990.
[293] Oxford, Oxford University Press, 1990.

in a position to assist others in this regard. The Committee notes in particular the importance of the Declaration on the Right to Development adopted by the General Assembly in its resolution 41/128 of 4 December 1986 and the need for States parties to take full account of all of the principles recognized therein. It emphasizes that, in the absence of an active programme of international assistance and cooperation on the part of all those States that are in a position to undertake one, the full realization of economic, social and cultural rights will remain an unfulfilled aspiration in many countries. In this respect, the Committee also recalls the terms of its General Comment 2 (1990).

General Comment 12

The right to adequate food

The right to adequate food (Art.11) 12/05/99. E/C.12/1999/5, CESCR General comment 12.

COMMITTEE ON ECONOMIC, SOCIAL AND CULTURAL RIGHTS
Twentieth session
Geneva, 26 April-14 May 1999

SUBSTANTIVE ISSUES ARISING IN THE IMPLEMENTATION OF THE INTERNATIONAL COVENANT ON ECONOMIC, SOCIAL AND CULTURAL RIGHTS:

(Art. 11)

Introduction and basic premises

1. The human right to adequate food is recognized in several instruments under international law. The International Covenant on Economic, Social and Cultural Rights deals more comprehensively than any other instrument with this right. Pursuant to Article 11.1 of the Covenant, States parties recognize 'the right of everyone to an adequate standard of living for himself and his family, including adequate food, clothing and housing, and to the continuous improvement of living conditions', while pursuant to Article 11.2 they recognize that more immediate and urgent steps may be needed to ensure 'the fundamental right to freedom from hunger and malnutrition'. The human right to adequate food is of crucial importance for the enjoyment of all rights. It applies to everyone; thus the reference in Article 11.1 to 'himself and his family' does not imply any limitation upon the applicability of this right to individuals or to female-headed households.
2. The Committee has accumulated significant information pertaining to the right to adequate food through examination of State parties' reports over the years since 1979. The Committee has noted that while reporting guidelines are available relating to the right to adequate food, only few States parties have provided information sufficient

and precise enough to enable the Committee to determine the prevailing situation in the countries concerned with respect to this right and to identify the obstacles to its realization. This General Comment aims to identify some of the principal issues which the Committee considers to be important in relation to the right to adequate food. Its preparation was triggered by the request of Member States during the 1996 World Food Summit, for a better definition of the rights relating to food in Article 11 of the Covenant, and by a special request to the Committee to give particular attention to the Summit Plan of Action in monitoring the implementation of the specific measures provided for in Article 11 of the Covenant.

3. In response to these requests, the Committee reviewed the relevant reports and documentation of the Commission on Human Rights and of the Sub-Commission on Prevention of Discrimination and Protection of Minorities on the right to adequate food as a human right; devoted a day of general discussion to this issue at its seventeenth session in 1997, taking into consideration the draft international code of conduct on the human right to adequate food prepared by international non-governmental organizations; participated in two expert consultations on the right to adequate food as a human right organized by the Office of the United Nations High Commissioner for Human Rights (OHCHR), in Geneva in December 1997, and in Rome in November 1998 co-hosted by the Food and Agriculture Organization of the United Nations (FAO), and noted their final reports. In April 1999 the Committee participated in a symposium on 'The substance and politics of a human rights approach to food and nutrition policies and programmes', organized by the Administrative Committee on Co-ordination/Sub-Committee on Nutrition of the United Nations at its twenty-sixth session in Geneva and hosted by OHCHR.

4. The Committee affirms that the right to adequate food is indivisibly linked to the inherent dignity of the human person and is indispensable for the fulfilment of other human rights enshrined in the International Bill of Human Rights. It is also inseparable from social justice, requiring the adoption of appropriate economic, environmental and social policies, at both the national and international levels, oriented to the eradication of poverty and the fulfilment of all human rights for all.

5. Despite the fact that the international community has frequently reaffirmed the importance of full respect for the right to adequate food, a disturbing gap still exists between the standards set in Article 11 of the Covenant and the situation prevailing in many parts of the world. More than 840 million people throughout the world, most of them in developing countries, are chronically hungry; millions of people are suffering from famine as the result of natural disasters, the increasing incidence of civil strife and wars in some regions and the use of food as a political weapon. The Committee observes that while the problems of hunger and malnutrition are often particularly acute in developing countries, malnutrition, under-nutrition and other problems which relate to the right to adequate food and the right to freedom from hunger, also exist in some of the most economically developed countries. Fundamentally, the roots of the problem of hunger and malnutrition are not lack of food but lack of access to available food, inter alia because of poverty, by large segments of the world's population.

Normative content of Article 11, paragraphs 1 and 2

6. The right to adequate food is realized when every man, woman and child, alone or in community with others, has physical and economic access at all times to adequate food or means for its procurement. The right to adequate food shall therefore not be interpreted in a narrow or restrictive sense which equates it with a minimum package of calories, proteins and other specific nutrients. The right to adequate food will have to be realized progressively. However, States have a core obligation to take the necessary action to mitigate and alleviate hunger as provided for in paragraph 2 of Article 11, even in times of natural or other disasters.

Adequacy and sustainability of food availability and access

7. The concept of adequacy is particularly significant in relation to the right to food since it serves to underline a number of factors which must be taken into account in determining whether particular foods or diets that are accessible can be considered the most appropriate under given circumstances for the purposes of Article 11 of the Covenant. The notion of sustainability is intrinsically linked to the notion of adequate food or food security, implying food being accessible for both present and future generations. The precise meaning of 'adequacy' is to a large extent determined by prevailing social, economic, cultural, climatic, ecological and other conditions, while 'sustainability' incorporates the notion of long-term availability and accessibility.
8. The Committee considers that the core content of the right to adequate food implies:
   - The availability of food in a quantity and quality sufficient to satisfy the dietary needs of individuals, free from adverse substances, and acceptable within a given culture;
   - The accessibility of such food in ways that are sustainable and that do not interfere with the enjoyment of other human rights.
9. Dietary needs implies that the diet as a whole contains a mix of nutrients for physical and mental growth, development and maintenance, and physical activity that are in compliance with human physiological needs at all stages throughout the life cycle and according to gender and occupation. Measures may therefore need to be taken to maintain, adapt or strengthen dietary diversity and appropriate consumption and feeding patterns, including breast-feeding, while ensuring that changes in availability and access to food supply as a minimum do not negatively affect dietary composition and intake.
10. Free from adverse substances sets requirements for food safety and for a range of protective measures by both public and private means to prevent contamination of foodstuffs through adulteration and/or through bad environmental hygiene or inappropriate handling at different stages throughout the food chain; care must also be taken to identify and avoid or destroy naturally occurring toxins.
11. Cultural or consumer acceptability implies the need also to take into account, as far as possible, perceived non nutrient-based values attached to food and food consumption and informed consumer concerns regarding the nature of accessible food supplies.

12. Availability refers to the possibilities either for feeding oneself directly from productive land or other natural resources, or for well functioning distribution, processing and market systems that can move food from the site of production to where it is needed in accordance with demand.
13. Accessibility encompasses both economic and physical accessibility:
    - Economic accessibility implies that personal or household financial costs associated with the acquisition of food for an adequate diet should be at a level such that the attainment and satisfaction of other basic needs are not threatened or compromised. Economic accessibility applies to any acquisition pattern or entitlement through which people procure their food and is a measure of the extent to which it is satisfactory for the enjoyment of the right to adequate food. Socially vulnerable groups such as landless persons and other particularly impoverished segments of the population may need attention through special programmes.
    - Physical accessibility implies that adequate food must be accessible to everyone, including physically vulnerable individuals, such as infants and young children, elderly people, the physically disabled, the terminally ill and persons with persistent medical problems, including the mentally ill. Victims of natural disasters, people living in disaster-prone areas and other specially disadvantaged groups may need special attention and sometimes priority consideration with respect to accessibility of food. A particular vulnerability is that of many indigenous population groups whose access to their ancestral lands may be threatened.

Obligations and violations

14. The nature of the legal obligations of States parties are set out in Article 2 of the Covenant and has been dealt with in the Committee's General Comment No. 3 (1990). The principal obligation is to take steps to achieve progressively the full realization of the right to adequate food. This imposes an obligation to move as expeditiously as possible towards that goal. Every State is obliged to ensure for everyone under its jurisdiction access to the minimum essential food which is sufficient, nutritionally adequate and safe, to ensure their freedom from hunger.
15. The right to adequate food, like any other human right, imposes three types or levels of obligations on States parties: the obligations to respect, to protect and to fulfil. In turn, the obligation to fulfil incorporates both an obligation to facilitate and an obligation to provide.[294] The obligation to respect existing access to adequate food requires States parties not to take any measures that result in preventing such access. The obligation to protect requires measures by the State to ensure that enterprises or individuals do not deprive individuals of their access to adequate food. The obligation to fulfil (facilitate) means the State must pro-actively engage in activities intended to

[294] Originally three levels of obligations were proposed: to respect, protect and assist/fulfil. (See Right to adequate food as a human right, Study Series No. 1, New York, 1989 (United Nations publication, Sales No. E.89.XIV.2).) The intermediate level of 'to facilitate' has been proposed as a Committee category, but the Committee decided to maintain the three levels of obligation.

strengthen people's access to and utilization of resources and means to ensure their livelihood, including food security. Finally, whenever an individual or group is unable, for reasons beyond their control, to enjoy the right to adequate food by the means at their disposal, States have the obligation to fulfil (provide) that right directly. This obligation also applies for persons who are victims of natural or other disasters.

16. Some measures at these different levels of obligations of States parties are of a more immediate nature, while other measures are more of a long-term character, to achieve progressively the full realization of the right to food.

17. Violations of the Covenant occur when a State fails to ensure the satisfaction of, at the very least, the minimum essential level required to be free from hunger. In determining which actions or omissions amount to a violation of the right to food, it is important to distinguish the inability from the unwillingness of a State party to comply. Should a State party argue that resource constraints make it impossible to provide access to food for those who are unable by themselves to secure such access, the State has to demonstrate that every effort has been made to use all the resources at its disposal in an effort to satisfy, as a matter of priority, those minimum obligations. This follows from Article 2.1 of the Covenant, which obliges a State party to take the necessary steps to the maximum of its available resources, as previously pointed out by the Committee in its General Comment No. 3, paragraph 10. A State claiming that it is unable to carry out its obligation for reasons beyond its control therefore has the burden of proving that this is the case and that it has unsuccessfully sought to obtain international support to ensure the availability and accessibility of the necessary food.

18. Furthermore, any discrimination in access to food, as well as to means and entitlements for its procurement, on the grounds of race, colour, sex, language, age, religion, political or other opinion, national or social origin, property, birth or other status with the purpose or effect of nullifying or impairing the equal enjoyment or exercise of economic, social and cultural rights constitutes a violation of the Covenant.

19. Violations of the right to food can occur through the direct action of States or other entities insufficiently regulated by States. These include: the formal repeal or suspension of legislation necessary for the continued enjoyment of the right to food; denial of access to food to particular individuals or groups, whether the discrimination is based on legislation or is pro-active; the prevention of access to humanitarian food aid in internal conflicts or other emergency situations; adoption of legislation or policies which are manifestly incompatible with pre-existing legal obligations relating to the right to food; and failure to regulate activities of individuals or groups so as to prevent them from violating the right to food of others, or the failure of a State to take into account its international legal obligations regarding the right to food when entering into agreements with other States or with international organizations.

20. While only States are parties to the Covenant and are thus ultimately accountable for compliance with it, all members of society – individuals, families, local communities, non-governmental organizations, civil society organizations, as well as the private business sector – have responsibilities in the realization of the right to adequate food. The State should provide an environment that facilitates implementation of these

responsibilities. The private business sector – national and transnational – should pursue its activities within the framework of a code of conduct conducive to respect of the right to adequate food, agreed upon jointly with the Government and civil society.

Implementation at the national level

21. The most appropriate ways and means of implementing the right to adequate food will inevitably vary significantly from one State party to another. Every State will have a margin of discretion in choosing its own approaches, but the Covenant clearly requires that each State party take whatever steps are necessary to ensure that everyone is free from hunger and as soon as possible can enjoy the right to adequate food. This will require the adoption of a national strategy to ensure food and nutrition security for all, based on human rights principles that define the objectives, and the formulation of policies and corresponding benchmarks. It should also identify the resources available to meet the objectives and the most cost-effective way of using them.
22. The strategy should be based on a systematic identification of policy measures and activities relevant to the situation and context, as derived from the normative content of the right to adequate food and spelled out in relation to the levels and nature of State parties' obligations referred to in paragraph 15 of the present general comment. This will facilitate coordination between ministries and regional and local authorities and ensure that related policies and administrative decisions are in compliance with the obligations under Article 11 of the Covenant.
23. The formulation and implementation of national strategies for the right to food requires full compliance with the principles of accountability, transparency, people's participation, decentralization, legislative capacity and the independence of the judiciary. Good governance is essential to the realization of all human rights, including the elimination of poverty and ensuring a satisfactory livelihood for all.
24. Appropriate institutional mechanisms should be devised to secure a representative process towards the formulation of a strategy, drawing on all available domestic expertise relevant to food and nutrition. The strategy should set out the responsibilities and time-frame for the implementation of the necessary measures.
25. The strategy should address critical issues and measures in regard to all aspects of the food system, including the production, processing, distribution, marketing and consumption of safe food, as well as parallel measures in the fields of health, education, employment and social security. Care should be taken to ensure the most sustainable management and use of natural and other resources for food at the national, regional, local and household levels.
26. The strategy should give particular attention to the need to prevent discrimination in access to food or resources for food. This should include: guarantees of full and equal access to economic resources, particularly for women, including the right to inheritance and the ownership of land and other property, credit, natural resources and appropriate technology; measures to respect and protect self-employment and work which provides a remuneration ensuring a decent living for wage earners and

their families (as stipulated in Article 7 (a) (ii) of the Covenant); maintaining registries on rights in land (including forests).

27. As part of their obligations to protect people's resource base for food, States parties should take appropriate steps to ensure that activities of the private business sector and civil society are in conformity with the right to food.
28. Even where a State faces severe resource constraints, whether caused by a process of economic adjustment, economic recession, climatic conditions or other factors, measures should be undertaken to ensure that the right to adequate food is especially fulfilled for vulnerable population groups and individuals.

Benchmarks and framework legislation

29. In implementing the country-specific strategies referred to above, States should set verifiable benchmarks for subsequent national and international monitoring. In this connection, States should consider the adoption of a framework law as a major instrument in the implementation of the national strategy concerning the right to food. The framework law should include provisions on its purpose; the targets or goals to be achieved and the time-frame to be set for the achievement of those targets; the means by which the purpose could be achieved described in broad terms, in particular the intended collaboration with civil society and the private sector and with international organizations; institutional responsibility for the process; and the national mechanisms for its monitoring, as well as possible recourse procedures. In developing the benchmarks and framework legislation, States parties should actively involve civil society organizations.
30. Appropriate United Nations programmes and agencies should assist, upon request, in drafting the framework legislation and in reviewing the sectoral legislation. FAO, for example, has considerable expertise and accumulated knowledge concerning legislation in the field of food and agriculture. The United Nations Children's Fund (UNICEF) has equivalent expertise concerning legislation with regard to the right to adequate food for infants and young children through maternal and child protection including legislation to enable breast-feeding, and with regard to the regulation of marketing of breast milk substitutes.

Monitoring

31. States parties shall develop and maintain mechanisms to monitor progress towards the realization of the right to adequate food for all, to identify the factors and difficulties affecting the degree of implementation of their obligations, and to facilitate the adoption of corrective legislation and administrative measures, including measures to implement their obligations under Articles 2.1 and 23 of the Covenant.

Remedies and accountability

32. Any person or group who is a victim of a violation of the right to adequate food should have access to effective judicial or other appropriate remedies at both national and international levels. All victims of such violations are entitled to adequate reparation, which may take the form of restitution, compensation, satisfaction or guarantees of

non-repetition. National Ombudsmen and human rights commissions should address violations of the right to food.

33. The incorporation in the domestic legal order of international instruments recognizing the right to food, or recognition of their applicability, can significantly enhance the scope and effectiveness of remedial measures and should be encouraged in all cases. Courts would then be empowered to adjudicate violations of the core content of the right to food by direct reference to obligations under the Covenant.
34. Judges and other members of the legal profession are invited to pay greater attention to violations of the right to food in the exercise of their functions.
35. States parties should respect and protect the work of human rights advocates and other members of civil society who assist vulnerable groups in the realization of their right to adequate food.

International obligations

States parties

36. In the spirit of Article 56 of the Charter of the United Nations, the specific provisions contained in Articles 11, 2.1, and 23 of the Covenant and the Rome Declaration of the World Food Summit, States parties should recognize the essential role of international cooperation and comply with their commitment to take joint and separate action to achieve the full realization of the right to adequate food. In implementing this commitment, States parties should take steps to respect the enjoyment of the right to food in other countries, to protect that right, to facilitate access to food and to provide the necessary aid when required. States parties should, in international agreements whenever relevant, ensure that the right to adequate food is given due attention and consider the development of further international legal instruments to that end.
37. States parties should refrain at all times from food embargoes or similar measures which endanger conditions for food production and access to food in other countries. Food should never be used as an instrument of political and economic pressure. In this regard, the Committee recalls its position, stated in its General Comment No. 8, on the relationship between economic sanctions and respect for economic, social and cultural rights.

States and international organizations

38. States have a joint and individual responsibility, in accordance with the Charter of the United Nations, to cooperate in providing disaster relief and humanitarian assistance in times of emergency, including assistance to refugees and internally displaced persons. Each State should contribute to this task in accordance with its ability. The role of the World Food Programme (WFP) and the Office of the United Nations High Commissioner for Refugees (UNHCR), and increasingly that of UNICEF and FAO is of particular importance in this respect and should be strengthened. Priority in food aid should be given to the most vulnerable populations.
39. Food aid should, as far as possible, be provided in ways which do not adversely affect local producers and local markets, and should be organized in ways that facilitate

the return to food self-reliance of the beneficiaries. Such aid should be based on the needs of the intended beneficiaries. Products included in international food trade or aid programmes must be safe and culturally acceptable to the recipient population.

The United Nations and other international organizations

40. The role of the United Nations agencies, including through the United Nations Development Assistance Framework (UNDAF) at the country level, in promoting the realization of the right to food is of special importance. Coordinated efforts for the realization of the right to food should be maintained to enhance coherence and interaction among all the actors concerned, including the various components of civil society. The food organizations, FAO, WFP and the International Fund for Agricultural Development (IFAD) in conjunction with the United Nations Development Programme (UNDP), UNICEF, the World Bank and the regional development banks, should cooperate more effectively, building on their respective expertise, on the implementation of the right to food at the national level, with due respect to their individual mandates.
41. The international financial institutions, notably the International Monetary Fund (IMF) and the World Bank, should pay greater attention to the protection of the right to food in their lending policies and credit agreements and in international measures to deal with the debt crisis. Care should be taken, in line with the Committee's General Comment No. 2, paragraph 9, in any structural adjustment programme to ensure that the right to food is protected.

Annex 3 Voluntary guidelines[295]

to support the progressive realization of the right to adequate food in the context of national food security

I Preface and introduction

Preface

1. The eradication of hunger is clearly reflected in the target set at the World Food Summit to reduce the number of undernourished people to half their present level no later than 2015 and as agreed by the Millennium Summit to 'halve the proportion of people who suffer from hunger' by the same year.
2. In the Rome Declaration on World Food Security, Heads of State and Government 'reaffirm[ed] the right of everyone to have access to safe and nutritious food, consistent with the right to adequate food and the fundamental right of everyone to be free from hunger.' Objective 7.4 of the World Food Summit Plan of Action established the task: 'to clarify the content of the right to adequate food and the fundamental right of everyone to be free from hunger, as stated in the International Covenant on Economic, Social and Cultural Rights and other relevant international and regional instruments, and to give particular attention to implementation and full and progressive realization of this right as a means of achieving food security for all.'
3. The Plan of Action 'invite[d] the UN High Commissioner for Human Rights, in consultation with relevant treaty bodies, and in collaboration with relevant specialized agencies and programmes of the UN system and appropriate intergovernmental mechanisms, to better define the rights related to food in Article 11 of the Covenant and to propose ways to implement and realize these rights as a means of achieving the commitments and objectives of the World Food Summit, taking into account the possibility of formulating voluntary guidelines for food security for all.'
4. In response to the invitation by the World Food Summit, and following several international consultations, the Committee on Economic, Social and Cultural Rights adopted General Comment 12, which provided its experts' views on the progressive realization of the right to adequate food.
5. In Paragraph 10 of the Declaration adopted at the 2002 World Food Summit: *five years later*, Heads of State and Government invited the Council of the Food and Agriculture Organization of the United Nations to establish at its 123rd session an Intergovernmental Working Group (IGWG), in the context of the World Food Summit follow-up, with the following mandate: 'to elaborate, with the participation of stakeholders, in a period of two years, a set of voluntary guidelines to support Member Nations' efforts to achieve the progressive realization of the right to adequate food in the context of national food security'.

[295] FAO has granted permission for a reprint of the FAO voluntary guidelines in this edited volume. The editors are grateful for this.

6. The objective of these Voluntary Guidelines is to provide practical guidance to States in their implementation of the progressive realization of the right to adequate food in the context of national food security, in order to achieve the goals of the Plan of Action of the World Food Summit. Relevant stakeholders could also benefit from such guidance.
7. The Voluntary Guidelines take into account a wide range of important considerations and principles, including equality and non-discrimination, participation and inclusion, accountability and rule of law, and the principle that all human rights are universal, indivisible, interrelated and interdependent. Food should not be used as a tool for political and economic pressure.
8. In developing these Voluntary Guidelines, the IGWG has benefited from the active participation of international organizations, non-governmental organizations (NGOs) and representatives of civil society. The implementation of these Guidelines, which is primarily the responsibility of States, should benefit from the contribution of all members of civil society at large, including NGOs and the private sector.
9. These Voluntary Guidelines are a human rights-based practical tool addressed to all States. They do not establish legally binding obligations for States or international organizations, nor is any provision in them to be interpreted as amending, modifying or otherwise impairing rights and obligations under national and international law. States are encouraged to apply these Voluntary Guidelines in developing their strategies, policies, programmes and activities, and should do so without discrimination of any kind, such as to race, colour, sex, language, religion, political or other opinion, national or social origin, property, birth or other status.

Introduction

Basic instruments

10. These Voluntary Guidelines have taken into account relevant international instruments,[296] in particular those instruments in which the progressive realization of the right of everyone to an adequate standard of living, including adequate food, is enshrined.

Universal Declaration of Human Rights, Article 25:

1. Everyone has the right to a standard of living adequate for the health and well-being of himself and of his family, including food, clothing, housing and medical care and necessary social services, and the right to security in the event of unemployment, sickness, disability, widowhood, old age or other lack of livelihood in circumstances beyond his control.

International Covenant on Economic, Social and Cultural Rights, Article 11:

1. The States Parties to the present Covenant recognize the right of everyone to an adequate standard of living for himself and his family, including adequate food,

[296] References in the Voluntary Guidelines to the International Covenant on Economic, Social and Cultural Rights and other international treaties do not prejudice the position of any State with respect to signature, ratification or accession to those instruments.

clothing and housing, and to the continuous improvement of living conditions. The States Parties will take appropriate steps to ensure the realization of this right, recognizing to this effect the essential importance of international cooperation based on free consent.

2. The States Parties to the present Covenant, recognizing the fundamental right of everyone to be free from hunger, shall take, individually and through international cooperation, the measures, including specific programmes, which are needed:
   (a) to improve methods of production, conservation and distribution of food by making full use of technical and scientific knowledge, by disseminating knowledge of the principles of nutrition and by developing or reforming agrarian systems in such a way as to achieve the most efficient development and utilization of natural resources;
   (b) taking into account the problems of both food-importing and food-exporting countries, to ensure an equitable distribution of world food supplies in relation to need.

International Covenant on Economic, Social and Cultural Rights, Article 2:

1. Each State Party to the present Covenant undertakes to take steps, individually and through international assistance and cooperation, especially economic and technical, to the maximum of its available resources, with a view to achieving progressively the full realization of the rights recognized in the present Covenant by all appropriate means, including particularly the adoption of legislative measures.
2. The States Parties to the present Covenant undertake to guarantee that the rights enunciated in the present Covenant will be exercised without discrimination of any kind as to race, colour, sex, language, religion, political or other opinion, national or social origin, property, birth or other status.

11. Among others, Articles 55 and 56 of the Charter of the United Nations are relevant to these Voluntary Guidelines.

UN Charter, Article 55

With a view to the creation of conditions of stability and well-being which are necessary for peaceful and friendly relations among nations based on respect for the principle of equal rights and self-determination of peoples, the United Nations shall promote:

a. higher standards of living, full employment, and conditions of economic and social progress and development;
b. solutions of international economic, social, health, and related problems; and international cultural and educational co-operation; and
c. universal respect for, and observance of, human rights and fundamental freedoms for all without distinction as to race, sex, language, or religion.

UN Charter, Article 56

All Members pledge themselves to take joint and separate action in co-operation with the Organization for the achievement of the purposes set forth in Article 55.

12. Other international instruments, including the Convention on the Rights of the Child and the Convention on the Elimination of All Forms of Discrimination Against

Women, the four Geneva Conventions and their two Additional Protocols also contain provisions relevant to these Voluntary Guidelines.

13. These Voluntary Guidelines have taken into account the commitments contained in the Millennium Declaration, including the development goals, as well as the outcomes and commitments of the major UN conferences and summits in the economic, social and related fields.
14. The IGWG has also taken into account several Resolutions from the United Nations General Assembly and Commission on Human Rights and the General Comments adopted by the Committee on Economic, Social and Cultural Rights.

The right to adequate food and the achievement of food security

15. Food security exists when all people, at all times, have physical and economic access to sufficient, safe and nutritious food to meet their dietary needs and food preferences for an active and healthy life. The four pillars of food security are availability, stability of supply, access and utilization.
16. The progressive realization of the right to adequate food requires States to fulfil their relevant human rights obligations under international law. These Voluntary Guidelines aim to guarantee the availability of food in quantity and quality sufficient to satisfy the dietary needs of individuals; physical and economic accessibility for everyone, including vulnerable groups, to adequate food, free from unsafe substances and acceptable within a given culture; or the means of its procurement.
17. States have obligations under relevant international instruments relevant to the progressive realization of the right to adequate food. Notably, States Parties to the International Covenant on Economic, Social and Cultural Rights (ICESCR) have the obligation to respect, promote and protect and to take appropriate steps to achieve progressively the full realization of the right to adequate food. States Parties should respect existing access to adequate food by not taking any measures that result in preventing such access, and should protect the right of everyone to adequate food by taking steps so that enterprises and individuals do not deprive individuals of their access to adequate food. States Parties should promote policies intended to contribute to the progressive realization of people's right to adequate food by proactively engaging in activities intended to strengthen people's access to and utilization of resources and means to ensure their livelihood, including food security. States Parties should, to the extent that resources permit, establish and maintain safety nets or other assistance to protect those who are unable to provide for themselves.
18. States that are not Parties to the International Covenant on Economic, Social and Cultural Rights (ICESCR) are invited to consider ratifying the ICESCR.
19. At the national level, a human rights-based approach to food security emphasizes universal, interdependent, indivisible and interrelated human rights, the obligations of States and the roles of relevant stakeholders. It emphasizes the achievement of food security as an outcome of the realization of existing rights and includes certain key principles: the need to enable individuals to realize the right to take part in the conduct of public affairs, the right to freedom of expression and the right to seek, receive and impart information, including in relation to decision-making about policies on realizing

the right to adequate food. Such an approach should take into account the need for emphasis on poor and vulnerable people who are often excluded from the processes that determine policies to promote food security and the need for inclusive societies free from discrimination by the State in meeting their obligations to promote and respect human rights. In this approach, people hold their governments accountable and are participants in the process of human development, rather than being passive recipients. A human rights-based approach requires not only addressing the final outcome of abolishing hunger, but also proposing ways and tools by which that goal is achieved. Application of human rights principles is integral to the process.

II Enabling environment, assistance and accountability

Guideline 1

Democracy, good governance, human rights and the rule of law

1.1 States should promote and safeguard a free, democratic and just society in order to provide a peaceful, stable and enabling economic, social, political and cultural environment in which individuals can feed themselves and their families in freedom and dignity.

1.2 States should promote democracy, the rule of law, sustainable development and good governance, and promote and protect human rights and fundamental freedoms in order to empower individuals and civil society to make demands on their governments, devise policies that address their specific needs and ensure the accountability and transparency of governments and state decision-making processes in implementing such policies. States should, in particular, promote freedom of opinion and expression, freedom of information, freedom of the press and freedom of assembly and association, which enhances the progressive realization of the right to adequate food in the context of national food security. Food should not be used as a tool for political and economic pressure.

1.3 States should also promote good governance as an essential factor for sustained economic growth, sustainable development, poverty and hunger eradication and for the realization of all human rights including the progressive realization of the right to adequate food.

1.4 States should ensure, in accordance with their international human rights obligations, that all individuals, including human rights defenders of the progressive realization of the right to adequate food, are accorded equal protection under the law and that due process is guaranteed in all legal proceedings.

1.5 Where appropriate and consistent with domestic law, States may assist individuals and groups of individuals to have access to legal assistance to better assert the progressive realization of the right to adequate food.

Guideline 2

Economic development policies

2.1 In order to achieve the progressive realization of the right to adequate food in the context of national food security, States should promote broad-based economic development that is supportive of their food security policies. States should establish policy goals and benchmarks based on the food security needs of their population.

2.2 States should assess, in consultation with key stakeholders, the economic and social situation, including the degree of food insecurity and its causes, the nutrition situation and food safety.

2.3 States should promote adequate and stable supplies of safe food through a combination of domestic production, trade, storage and distribution.

2.4 States should consider adopting a holistic and comprehensive approach to hunger and poverty reduction. Such an approach entails, inter alia, direct and immediate measures to ensure access to adequate food as part of a social safety net; investment in productive activities and projects to improve the livelihoods of the poor and hungry in a sustainable manner; the development of appropriate institutions, functioning markets, a conducive legal and regulatory framework; and access to employment, productive resources and appropriate services.

2.5 States should pursue inclusive, non-discriminatory and sound economic, agriculture, fisheries, forestry, land-use, and, as appropriate, land-reform policies, all of which will permit farmers, fishers, foresters and other food producers, particularly women, to earn a fair return from their labour, capital and management, and encourage conservation and sustainable management of natural resources, including in marginal areas.

2.6 Where poverty and hunger are predominantly rural, States should focus on sustainable agricultural and rural development through measures to improve access to land, water, appropriate and affordable technologies, productive and financial resources, enhance the productivity of poor rural communities, promote the participation of the poor in economic policy decisions, share the benefits of productivity gains, conserve and protect natural resources, and invest in rural infrastructure, education and research. In particular, States should adopt policies that create conditions that encourage stable employment, especially in rural areas, including off-farm jobs.

2.7 In response to the growing problem of urban hunger and poverty, States should promote investments aimed at enhancing the livelihoods of the urban poor.

Guideline 3

Strategies

3.1 States, as appropriate and in consultation with relevant stakeholders and pursuant to their national laws, should consider adopting a national human-rights based strategy for the progressive realization of the right to adequate food in the context of national food security as part of an overarching national development strategy, including poverty reduction strategies, where they exist.

3.2 The elaboration of these strategies should begin with a careful assessment of existing national legislation, policy and administrative measures, current programmes, systematic identification of existing constraints and availability of existing resources. States should formulate the measures necessary to remedy any weakness, and propose an agenda for change and the means for its implementation and evaluation.

3.3 These strategies could include objectives, targets, benchmarks and time frames; and actions to formulate policies, identify and mobilize resources, define institutional mechanisms, allocate responsibilities, coordinate the activities of different actors, and provide for monitoring mechanisms. As appropriate, such strategies could address all aspects of the food system, including the production, processing, distribution, marketing and consumption of safe food. They could also address access to resources and to markets as well as parallel measures in other fields. These strategies should, in particular, address the needs of vulnerable and disadvantaged groups, as well as special situations such as natural disasters and emergencies.

3.4 Where necessary, States should consider adopting and, as appropriate, reviewing a national poverty reduction strategy that specifically addresses access to adequate food.

3.5 States, individually or in cooperation with relevant international organizations, should consider integrating into their poverty reduction strategy a human rights perspective based on the principle of non-discrimination. In raising the standard of living of those below the poverty line, due regard should be given to the need to ensure equality in practice to those who are traditionally disadvantaged and between women and men.

3.6 In their poverty reduction strategies, States should also give priority to providing basic services for the poorest, and investing in human resources by ensuring access to primary education for all, basic health care, capacity building in good practices, clean drinking-water, adequate sanitation and justice and by supporting programmes in basic literacy, numeracy and good hygiene practices.

3.7 States are encouraged, inter alia and in a sustainable manner, to increase productivity and to revitalize the agriculture sector including livestock, forestry and fisheries through special policies and strategies targeted at small-scale and traditional fishers and farmers in rural areas, and the creation of enabling conditions for private sector participation, with emphasis on human capacity development and the removal of constraints to agricultural production, marketing and distribution.

3.8 In developing these strategies, States are encouraged to consult with civil society organizations and other key stakeholders at national and regional levels, including small-scale and traditional farmers, the private sector, women and youth associations, with the aim of promoting their active participation in all aspects of agricultural and food production strategies.

3.9 These strategies should be transparent, inclusive and comprehensive, cut across national policies, programmes and projects, take into account the special needs of girls and women, combine short-term and long-term objectives, and be prepared and implemented in a participatory and accountable manner.

3.10 States should support, including through regional cooperation, the implementation of national strategies for development, in particular for the reduction of poverty and hunger as well as for the progressive realization of the right to adequate food.

Guideline 4

Market systems

4.1 States should, in accordance with their national law and priorities, as well as their international commitments, improve the functioning of their markets, in particular their agricultural and food markets, in order to promote both economic growth and sustainable development, inter alia, by mobilizing domestic savings, both public and private, by developing appropriate credit policies, by generating sustainable adequate levels of national productive investment through credits in concessional terms and by increasing human capacity.

4.2 States should put legislation, policies, procedures and regulatory and other institutions in place to ensure non-discriminatory access to markets and to prevent uncompetitive practices in markets.

4.3 States should encourage the development of corporate social responsibility and the commitment of all market players and civil society towards the progressive realization of the right of individuals to adequate food in the context of national food security.

4.4 States should provide adequate protection to consumers against fraudulent market practices, misinformation and unsafe food. The measures toward this objective should not constitute unjustified barriers to international trade and should be in conformity with the WTO agreements.

4.5 States should, as appropriate, promote the development of small-scale local and regional markets and border trade to reduce poverty and increase food security, particularly in poor rural and urban areas.

4.6 States may wish to adopt measures to ensure that the widest number of individuals and communities, especially disadvantaged groups, can benefit from opportunities created by competitive agricultural trade.

4.7 States should strive to ensure that food, agricultural trade and overall trade policies are conducive to fostering food security for all through a non-discriminatory and market-oriented local, regional, national and world trade system.

4.8 States should endeavour to establish well functioning internal marketing, storage, transportation, communication and distribution systems, inter alia, to facilitate diversified trade and better links within and between domestic, regional and world markets, as well as to take advantage of new market opportunities.

4.9 States will take into account that markets do not automatically result in everybody achieving a sufficient income at all times to meet basic needs, and should therefore seek to provide adequate social safety nets and, where appropriate, the assistance of the international community for this purpose.

4.10 States should take into account the shortcomings of market mechanisms in protecting the environment and public goods.

Guideline 5

Institutions

5.1 States, where appropriate, should assess the mandate and performance of relevant public institutions and, where necessary, establish, reform or improve their organization and structure to contribute to the progressive realization of the right to adequate food in the context of national food security.

5.2 To this end, States may wish to ensure the coordinated efforts of relevant government ministries, agencies and offices. They could establish national intersectoral coordination mechanisms to ensure the concerted implementation, monitoring and evaluation of policies, plans and programmes. States are encouraged to involve relevant communities in all aspects of planning and execution of activities in these areas.

5.3 States may also wish to entrust a specific institution with overall responsibility for overseeing and coordinating the application of these guidelines, bearing in mind the Declaration and Programme of Action of the 1993 Vienna World Conference on Human Rights and taking due account of existing agriculture conventions and protocols. In order to ensure transparency and accountability, the functions and tasks of this institution would need to be clearly defined, regularly reviewed and provision made for adequate monitoring mechanisms.

5.4 States should ensure that relevant institutions provide for full and transparent participation of the private sector and of civil society, in particular representatives of the groups most affected by food insecurity.

5.5 States should take measures, where and if necessary, to develop, strengthen, implement and maintain effective anticorruption legislation and policies, including in the food sector and in the management of emergency food aid.

Guideline 6

Stakeholders

6.1 Recognizing the primary responsibility of States for the progressive realization of the right to adequate food, States are encouraged to apply a multistakeholder approach to national food security to identify the roles of and involve all relevant stakeholders, encompassing civil society and the private sector, drawing together their know-how with a view to facilitating the efficient use of resources.

Guideline 7

Legal framework

7.1 States are invited to consider, in accordance with their domestic legal and policy frameworks, whether to include provisions in their domestic law, possibly including constitutional or legislative review that facilitates the progressive realization of the right to adequate food in the context of national food security.

7.2 States are invited to consider, in accordance with their domestic legal and policy frameworks, whether to include provisions in their domestic law, which may include their constitutions, bills of rights or legislation, to directly implement the progressive realization of the right to adequate food. Administrative, quasi-judicial and judicial mechanisms to provide adequate, effective and prompt remedies accessible, in particular, to members of vulnerable groups may be envisaged.

7.3 States that have established a right to adequate food under their legal system should inform the general public of all available rights and remedies to which they are entitled.

7.4 States should consider strengthening their domestic law and policies to accord access by women heads of households to poverty reduction and nutrition security programmes and projects.

Guideline 8

Access to resources and assets

8.1 States should facilitate sustainable, non-discriminatory and secure access and utilization of resources consistent with their national law and with international law and protect the assets that are important for people's livelihoods. States should respect and protect the rights of individuals with respect to resources such as land, water, forests, fisheries and livestock without any discrimination. Where necessary and appropriate, States should carry out land reforms and other policy reforms consistent with their human rights obligations and in accordance with the rule of law in order to secure efficient and equitable access to land and to strengthen propoor growth. Special attention may be given to groups such as pastoralists and indigenous people and their relation to natural resources.

8.2 States should take steps so that members of vulnerable groups can have access to opportunities and economic resources in order to participate fully and equally in the economy.

8.3 States should pay particular attention to the specific access problems of women and of vulnerable, marginalized and traditionally disadvantaged groups, including all persons affected by HIV/AIDS. States should take measures to protect all people affected by HIV/AIDS from losing their access to resources and assets.

8.4 States should promote agricultural research and development, in particular to promote basic food production with its positive effects on basic incomes and its benefits to small and women farmers, as well as poor consumers.

8.5 States should, within the framework of relevant international agreements, including those on intellectual property, promote access by medium- and smallscale farmers to research results enhancing food security.

8.6 States should promote women's full and equal participation in the economy and, for this purpose, introduce, where it does not exist, and implement gendersensitive legislation providing women with the right to inherit and possess land and other property. States should also provide women with secure and equal access to, control over, and benefits from productive resources, including credit, land, water and appropriate technologies.

8.7 States should design and implement programmes that include different mechanisms of access and appropriate use of agricultural land directed to the poorest populations.

Guideline 8A

Labour

8.8 States should take measures to encourage sustainable development in order to provide opportunities for work that provide remuneration allowing for an adequate standard of living for rural and urban wage earners and their families, and to promote and protect self-employment. For States that have ratified the relevant instruments, working conditions should be consistent with the obligations they have assumed under the International Covenant on Economic, Social and Cultural Rights, relevant ILO Conventions and other treaties including human rights conventions.

8.9 In order to improve access to the labour market, States should enhance human capital through education programmes, adult literacy and additional training programmes, as required, regardless of race, colour, gender, language, religion, political opinion, national or social origin, property, birth or other status.

Guideline 8B

Land

8.10 States should take measures to promote and protect the security of land tenure, especially with respect to women, and poor and disadvantaged segments of society, through legislation that protects the full and equal right to own land and other property, including the right to inherit. As appropriate, States should consider establishing legal and other policy mechanisms, consistent with their international human rights obligations and in accordance with the rule of law, that advance land reform to enhance access for the poor and women. Such mechanisms should also promote conservation and sustainable use of land. Special consideration should be given to the situation of indigenous communities.

Guideline 8C

Water

8.11 Bearing in mind that access to water in sufficient quantity and quality for all is fundamental for life and health, States should strive to improve access to, and promote sustainable use of, water resources and their allocation among users giving due regard to efficiency and the satisfaction of basic human needs in an equitable manner and that balances the requirement of preserving or restoring the functioning of ecosystems with domestic, industrial and agricultural needs, including safeguarding drinking-water quality.

Guideline 8D

Genetic resources for food and agriculture

8.12 States, taking into account the importance of biodiversity, and consistent with their obligations under relevant international agreements, should consider specific national policies, legal instruments and supporting mechanisms to prevent the erosion of and ensure the conservation and sustainable use of genetic resources for food and agriculture, including, as appropriate, for the protection of relevant traditional knowledge and equitable participation in sharing benefits arising from the use of these resources, and by encouraging, as appropriate, the participation of local and indigenous communities and farmers in making national decisions on matters related to the conservation and sustainable use of genetic resources for food and agriculture.

Guideline 8E

Sustainability

8.13 States should consider specific national policies, legal instruments and supporting mechanisms to protect ecological sustainability and the carrying capacity of ecosystems to ensure the possibility for increased, sustainable food production for present and future generations, prevent water pollution, protect the fertility of the soil, and promote the sustainable management of fisheries and forestry.

Guideline 8F

Services

8.14 States should create an enabling environment and strategies to facilitate and support the development of private and public sector initiatives to promote appropriate tools, technologies and mechanization in the provision of relevant services, including research, extension, marketing, rural finance and microcredit, to enable more efficient food production by all farmers, in particular poor farmers, and to address local constraints such as shortage of land, water and farm power.

Guideline 9

Food safety and consumer protection

9.1 States should take measures to ensure that all food, whether locally produced or imported, freely available or sold on markets, is safe and consistent with national food safety standards.

9.2 States should establish comprehensive and rational food-control systems that reduce risk of food-borne disease using risk analysis and supervisory mechanisms to ensure food safety in the entire food chain including animal feed.

9.3 States are encouraged to take action to streamline institutional procedures for food control and food safety at national level and eliminate gaps and overlaps in inspection systems and in the legislative and regulatory framework for food. States are encouraged to adopt scientifically based food safety standards, including standards for additives, contaminants, residues of veterinary drugs and pesticides, and microbiological hazards, and to establish standards for the packaging, labelling and advertising of food. These standards should take into consideration internationally accepted food standards (Codex Alimentarius) in accordance with the WTO Sanitary and Phytosanitary Agreement

(SPS). States should take action to prevent contamination from industrial and other pollutants in the production, processing, storage, transport, distribution, handling and sale of food.

9.4 States may wish to establish a national coordinating committee for food to bring together both governmental and non-governmental actors involved in the food system and to act as liaison with the FAO/WHO Codex Alimentarius Commission. States should consider collaborating with private stakeholders in the food system, both by assisting them in exercising controls on their own production and handling practices, and by auditing those controls.

9.5 Where necessary, States should assist farmers and other primary producers to follow good agricultural practices, food processors to follow good manufacturing practices, and food handlers to follow good hygiene practices. States are encouraged to consider establishing food safety systems and supervisory mechanisms to ensure the provision of safe food to consumers.

9.6 States should ensure that education on safe practices is available for food business operators so that their activities neither lead to harmful residues in food nor cause harm to the environment. States should also take measures to educate consumers about the safe storage, handling and utilization of food within the household. States should collect and disseminate information to the public regarding food-borne diseases and food safety matters, and should cooperate with regional and international organizations addressing food safety issues.

9.7 States should adopt measures to protect consumers from deception and misrepresentation in the packaging, labelling, advertising and sale of food and facilitate consumers' choice by ensuring appropriate information on marketed food, and provide recourse for any harm caused by unsafe or adulterated food, including food offered by street sellers. Such measures should not be used as unjustified barriers to trade; they should be in conformity with the WTO agreements (in particular SPS and TBT).

9.8 Developed countries are encouraged to provide technical assistance to developing countries through advice, credits, donations and grants for capacity building and training in food safety. When possible and appropriate, developing countries with more advanced capabilities in food safety-related areas are encouraged to lend assistance to less advanced developing countries.

9.9 States are encouraged to cooperate with all stakeholders, including regional and international consumer organizations, in addressing food safety issues, and consider their participation in national and international fora where policies with impact on food production, processing, distribution, storage and marketing are discussed.

Guideline 10

Nutrition

10.1 If necessary, States should take measures to maintain, adapt or strengthen dietary diversity and healthy eating habits and food preparation, as well as feeding patterns, including breastfeeding, while ensuring that changes in availability and access to food supply do not negatively affect dietary composition and intake.

10.2 States are encouraged to take steps, in particular through education, information and labelling regulations, to prevent overconsumption and unbalanced diets that may lead to malnutrition, obesity and degenerative diseases.

10.3 States are encouraged to involve all relevant stakeholders, in particular communities and local government, in the design, implementation, management, monitoring and evaluation of programmes to increase the production and consumption of healthy and nutritious foods, especially those that are rich in micronutrients. States may wish to promote gardens both at home and at school as a key element in combating micronutrient deficiencies and promoting healthy eating. States may also consider adopting regulations for fortifying foods to prevent and cure micronutrient deficiencies, in particular of iodine, iron and Vitamin A.

10.4 States should address the specific food and nutritional needs of people living with HIV/AIDS or suffering from other epidemics.

10.5 States should take appropriate measures to promote and encourage breastfeeding, in line with their cultures, the International Code of Marketing of Breast-milk Substitutes and subsequent resolutions of the World Health Assembly, in accordance with the WHO/UNICEF recommendations.

10.6 States may wish to disseminate information on the feeding of infants and young children that is consistent and in line with current scientific knowledge and internationally accepted practices and to take steps to counteract misinformation on infant feeding. States should consider with utmost care issues regarding breastfeeding and human immunodeficiency virus (HIV) infection on the basis of the most up-to-date, authoritative scientific advice and referring to the latest WHO/UNICEF guidelines.

10.7 States are invited to take parallel action in the areas of health, education and sanitary infrastructure and promote intersectoral collaboration, so that necessary services and goods become available to people to enable them to make full use of the dietary value in the food they eat and thus achieve nutritional well-being.

10.8 States should adopt measures to eradicate any kind of discriminatory practices, especially with respect to gender, in order to achieve adequate levels of nutrition within the household.

10.9 States should recognize that food is a vital part of an individual's culture, and they are encouraged to take into account individuals' practices, customs and traditions on matters related to food.

10.10 States are reminded of the cultural values of dietary and eating habits in different cultures and should establish methods for promoting food safety, positive nutritional intake including fair distribution of food within communities and households with special emphasis on the needs and rights of both girls and boys, as well as pregnant women and lactating mothers, in all cultures.

Guideline 11

Education and awareness raising

11.1 States should support investment in human resource development such as health, education, literacy and other skills training, which are essential to sustainable development, including agriculture, fisheries, forestry and rural development.

11.2 States should strengthen and broaden primary education opportunities, especially for girls, women and other underserved populations.

11.3 States should encourage agricultural and environmental education at the primary and secondary levels in order to create a better awareness in new generations about the importance of conserving and making sustainable use of natural resources.

11.4 States should support higher education by strengthening developing country university and technical faculties of agriculture-related disciplines and business to carry out both education and research functions, and by engaging universities throughout the world in training developing country agriculturalists, scientists and businesspeople at the graduate and post-graduate levels.

11.5 States should provide information to individuals to strengthen their ability to participate in food-related policy decisions that may affect them, and to challenge decisions that threaten their rights.

11.6 States should implement measures to make people improve their housing conditions and their means for food preparation, because they are related to food safety. Such measures should be made in the educative and infrastructure fields, especially in rural households.

11.7 States should promote, and/or integrate into school curricula, human rights education, including civil, political, economic, social and cultural rights, which includes the progressive realization of the right to adequate food.

11.8 States are encouraged to promote awareness of the importance of human rights, including the progressive realization of the right to adequate food.

11.9 States should provide proper training to officials responsible for the implementation of the progressive realization of the right to adequate food.

11.10 States should raise public awareness of these guidelines and continuously provide and improve access to them and to relevant human rights laws and regulations, particularly in rural and remote areas.

11.11 States may wish to empower civil society to participate in the implementation of these guidelines, for instance through capacity building.

Guideline 12

National financial resources

12.1 Regional and local authorities are encouraged to allocate resources for antihunger and food security purposes in their respective budgets.

12.2 States should ensure transparency and accountability in the use of public resources, particularly in the area of food security.

12.3 States are encouraged to promote basic social programmes and expenditures, in particular those affecting the poor and the vulnerable segments of society, and to protect them from budget reductions, while increasing the quality and effectiveness of social expenditures. States should strive to ensure that budget cuts do not negatively affect access to adequate food among the poorest sections of society.

12.4 States are encouraged to establish an enabling legal and economic environment to promote and mobilize domestic savings and attract external resources for productive investment, and seek innovative sources of funding, both public and private at national and international levels, for social programmes.

12.5 States are invited to take appropriate steps and suggest strategies to contribute to raise awareness of the families of migrants in order to promote efficient use of the remittances of migrants for investments that could improve their livelihoods, including the food security of their families.

Guideline 13

Support for vulnerable groups

13.1 Consistent with the World Food Summit commitment, States should establish Food Insecurity and Vulnerability Information and Mapping Systems (FIVIMS), in order to identify groups and households particularly vulnerable to food insecurity along with the reasons for their food insecurity. States should develop and identify corrective measures to be implemented both immediately and progressively to provide access to adequate food.

13.2 States are invited to systematically undertake disaggregated analysis on the food insecurity, vulnerability and nutritional status of different groups in society, with particular attention to assessing any form of discrimination that may manifest itself in greater food insecurity and vulnerability to food insecurity, or in a higher prevalence of malnutrition among specific population groups, or both, with a view to removing and preventing such causes of food insecurity or malnutrition.

13.3 States should establish transparent, non-discriminatory eligibility criteria in order to ensure effective targeting of assistance, so that no one who is in need is excluded, or that those not in need of assistance are included. Effective accountability and administrative systems are essential to prevent leakages and corruption. Factors to take into account include household and individual assets and income, nutrition and health status, as well as existing coping mechanisms.

13.4 States may wish to give priority to channelling food assistance via women as a means of enhancing their decision-making role and ensuring that the food is used to meet the household's food requirements.

Guideline 14

Safety nets

14.1 States should consider, to the extent that resources permit, establishing and maintaining social safety and food safety nets to protect those who are unable to provide for themselves. As far as possible, and with due regard to effectiveness and coverage, States should consider building on existing capacities within communities at risk to provide the necessary resources for social safety and food safety nets to fulfil the progressive realization of the right to adequate food. States may wish to consider the benefits of procuring locally.

14.2 States and international organizations should consider the benefits of local procurement for food assistance that could integrate the nutritional needs of those affected by food insecurity and the commercial interests of local producers.

14.3 Although the design of social and food safety nets will depend on the nature of food insecurity, objectives, budget, existing administrative capacity and local circumstances such as levels of food supply and local food markets, States should nonetheless ensure that they adequately target those in need and respect the principle of non-discrimination in the establishment of eligibility criteria.

14.4 States should take steps, to the extent that resources permit, so that any measure of an economic or financial nature likely to have a negative impact on existing levels of food consumption of vulnerable groups be accompanied by provision for effective food safety nets. Safety nets should be linked to other complementary interventions that promote food security in the longer term.

14.5 In situations where it has been determined that food plays an appropriate role in safety nets, food assistance should bridge the gap between the nutritional needs of the affected population and their ability to meet those needs themselves. Food assistance should be provided with the fullest possible participation of those affected, and such food should be nutritionally adequate and safe, bearing in mind local circumstances, dietary traditions and cultures.

14.6 States should consider accompanying food assistance in safety net schemes with complementary activities to maximize benefits towards ensuring people's access to and utilization of adequate food. Essential complementary activities include access to clean water and sanitation, health care interventions and nutrition education activities.

14.7 States, in the design of safety nets, should consider the important role of international organizations such as FAO, IFAD and WFP, and other relevant international, regional and civil society organizations that can assist them in fighting rural poverty and promoting food security and agricultural development.

Guideline 15

International food aid

15.1 Donor States should ensure that their food aid policies support national efforts by recipient States to achieve food security, and base their food aid provisions on sound needs assessment, targeting especially food insecure and vulnerable groups. In this context, donor States should provide assistance in a manner that takes into account food safety, the importance of not disrupting local food production and the nutritional and dietary needs and cultures of recipient populations. Food aid should be provided with a clear exit strategy and avoid the creation of dependency. Donors should promote increased use of local and regional commercial markets to meet food needs in famine-prone countries and reduce dependence on food aid.

15.2 International food-aid transactions, including bilateral food aid that is monetized, should be carried out in a manner consistent with the FAO Principles of Surplus Disposal and Consultative Obligations, the Food Aid Convention and the WTO Agreement on Agriculture, and should meet the internationally agreed food safety standards, bearing in mind local circumstances, dietary traditions and cultures.

15.3 States and relevant non-state actors should ensure, in accordance with international law, safe and unimpeded access to the populations in need, as well as for international needs assessments, and by humanitarian agencies involved in the distribution of international food assistance.

15.4 The provision of international food aid in emergency situations should take particular account of longer-term rehabilitation and development objectives in the recipient countries, and should respect universally recognized humanitarian principles.

15.5 The assessment of needs and the planning, monitoring and evaluation of the provision of food aid should, as far as possible, be made in a participatory manner and, whenever possible, in close collaboration with recipient governments at the national and local level.

Guideline 16

Natural and human-made disasters

16.1 Food should never be used as a means of political and economic pressure.

16.2 States reaffirm the obligations they have assumed under international humanitarian law and, in particular, as parties to the 1949 Geneva Conventions and/or the 1977 Additional Protocols thereto with respect to the humanitarian needs of the civilian population, including their access to food in situations of armed conflict and occupation, inter alia, Additional Protocol I provides, inter alia, that '[t]he starvation of civilians as a method of warfare is prohibited' and that '[i]t is prohibited to attack, destroy, remove or render useless objects indispensable to the survival of the civilian population, such as foodstuffs, agricultural areas for the production of foodstuffs, crops, livestock, drinking water installations and supplies and irrigation works, for the specific purpose of denying them, for their sustenance value to the civilian population or to the adverse party, whatever the motive, whether in order to starve out civilians, to cause them to move away, or for any other motive', and that 'these objects shall not be made the object of reprisals'.

16.3 In situations of occupation, international humanitarian law provides, inter alia: that to the fullest extent of the means available to it, the Occupying Power has the duty of ensuring the food and medical supplies of the population; that it should, in particular, bring in the necessary foodstuffs, medical stores and other articles if the resources of the Occupied Territory are inadequate; and that if the whole or part of the population of an Occupied Territory is inadequately supplied, the Occupying Power shall agree to relief schemes on behalf of the said population, and shall facilitate them by all the means at its disposal.[297]

16.4 States reaffirm the obligations they have assumed regarding the protection, safety and security of humanitarian personnel.

16.5 States should make every effort to ensure that refugees and internally displaced persons have access at all times to adequate food. In this respect, States and other relevant stakeholders should be encouraged to make use of the Guiding Principles on Internal Displacement when dealing with situations of internal displacement.

16.6 In the case of natural or human-made disasters, States should provide food assistance to those in need, may request international assistance if their own resources do not suffice, and should facilitate safe and unimpeded access for international assistance in accordance with international law and universally recognized humanitarian principles, bearing in mind local circumstances, dietary traditions and cultures.

16.7 States should put in place adequate and functioning mechanisms of early warning to prevent or mitigate the effects of natural or human-made disasters. Early warning systems should be based on international standards and cooperation, on reliable, disaggregated data and should be constantly monitored. States should take appropriate emergency preparedness measures, such as keeping food stocks for the acquisition of food, and take steps to put in place adequate systems for distribution.

16.8 States are invited to consider establishing mechanisms to assess nutritional impact and to gain understanding of the coping strategies of affected households in the event of natural or human-made disasters. This should inform the targeting, design, implementation and evaluation of relief, rehabilitation and resilience building programmes.

Guideline 17

Monitoring, indicators and benchmarks

17.1 States may wish to establish mechanisms to monitor and evaluate the implementation of these guidelines towards the progressive realization of the right to adequate food in the context of national food security, in accordance with their capacity and by building on existing information systems and addressing information gaps.

17.2 States may wish to consider conducting 'Right to Food Impact Assessments' in order to identify the impact of domestic policies, programmes and projects on the progressive realization of the right to adequate food of the population at large and vulnerable groups in particular, and as a basis for the adoption of the necessary corrective measures.

17.3 States may also wish to develop a set of process, impact and outcome indicators, relying on indicators already in use and monitoring systems such as FIVIMS, so as to assess the implementation of the progressive realization of the right to adequate food. They may wish to establish appropriate benchmarks to be achieved in the

[297] 1949 Geneva Convention IV Relative to the Protection of Civilian Persons in Time of War, Articles 55, 59.

short, medium and long term, which relate directly to meeting poverty and hunger reduction targets as a minimum, as well as other national and international goals including those adopted at the World Food Summit and the Millennium Summit.

17.4 In this evaluation process, process indicators could be so identified or designed that they explicitly relate and reflect the use of specific policy instruments and interventions with outcomes consistent with the progressive realization of the right to adequate food in the context of national food security. Such indicators could enable States to implement legal, policy and administrative measures, detect discriminatory practices and outcomes, and ascertain the extent of political and social participation in the process of realizing that right.

17.5 States should, in particular, monitor the food security situation of vulnerable groups, especially women, children and the elderly, and their nutritional status, including the prevalence of micronutrient deficiencies.

17.6 In this evaluation process, States should ensure a participatory approach to information gathering, management, analysis, interpretation and dissemination.

Guideline 18

National human rights institutions

18.1 States that have as a matter of national law or policy adopted a rights-based approach, and national human rights institutions or ombudspersons, may wish to include the progressive realization of the right to adequate food in the context of national food security in their mandates. States that do not have national human rights institutions or ombudspersons are encouraged to establish them. Human rights institutions should be independent and autonomous from the government, in accordance with the Paris Principles. States should encourage civil society organizations and individuals to contribute to monitoring activities undertaken by national human rights institutions with respect to the progressive realization of the right to adequate food.

18.2 States are invited to encourage efforts by national institutions to establish partnerships and increase cooperation with civil society.

Guideline 19

International dimension

19.1 States should fulfil those measures, actions and commitments on the international dimension, as described in Section III below, in support of the implementation of the Voluntary Guidelines, which assist States in their national efforts in the progressive realization of the right to adequate food in the context of national food security as set forth by the World Food Summit and the World Food Summit: five years later within the context of the Millennium Declaration.

III International measures, actions and commitments

International cooperation and unilateral measures

1. In the context of recent major international conferences, the international community has stated its deep concern over the persistence of hunger, its readiness to support national governments in their efforts to combat hunger and malnutrition and its commitment to cooperate actively within the global partnership for development, which includes the International Alliance Against Hunger.
2. States have the primary responsibility for their own economic and social development, including the progressive realization of the right to adequate food in the context of national food security. Stressing that national development efforts should be supported by an enabling international environment, the international community and the UN system, including FAO, as well as other relevant agencies and bodies according to their mandates, are urged to take actions in supporting national development efforts for the progressive realization of the right to adequate food in the context of national food security. This essential role of international cooperation is recognized, inter alia, in Article 56 of the Charter of the United Nations as well as in the outcomes of major international conferences such as the Plan of Implementation of the World Summit on Sustainable Development. Food should not be used as a tool of economic and political pressure.
3. States are strongly urged to take steps with a view to the avoidance of, and refrain from, any unilateral measure not in accordance with international law and the Charter of the United Nations that impedes the full achievement of economic and social development by the populations of the affected countries and that hinders their progressive realization of the right to adequate food.

Role of the international community

4. Consistent with commitments made at various international conferences, in particular the Monterrey Consensus, developed countries should assist developing countries in attaining international development goals, including those contained in the Millennium Declaration. States and relevant international organizations according to their respective mandates should actively support the progressive realization of the right to adequate food at the national level. External support, including South–South cooperation, should be coordinated with national policies and priorities.

Technical cooperation

5. Developed and developing countries should act in partnership to support their efforts to achieve the progressive realization of the right to adequate food in the context of national food security through technical cooperation, including institutional capacity building, and transfer of technology on mutually agreed terms, as committed in the

major international conferences, in all areas covered in these guidelines, with special focus on impediments to food security such as HIV/AIDS.

International trade

6. International trade can play a major role in the promotion of economic development, and the alleviation of poverty and improving food security at the national level.
7. States should promote international trade as one of the effective instruments for development, as expanded international trade could open opportunities to reduce hunger and poverty in many of the developing countries.
8. It is recalled that the long-term objective referred to in the WTO Agreement on Agriculture is to establish a fair and market-oriented trading system through a programme of fundamental reform encompassing strengthened rules and specific commitments on support and protection in order to correct and prevent restrictions and distortions in world agricultural markets.
9. States are urged to implement commitments expressed at various relevant international conferences and the recommendations of the São Paulo Consensus (the eleventh session of the United Nations Conference on Trade and Development) including, for example, those reproduced below:
   75. Agriculture is a central element in the current negotiations. Efforts should be intensified to achieve the internationally agreed aims embodied in the three pillars of the Doha mandate, namely substantial improvements in market access; reductions of, with a view to phasing out, all forms of export subsidies; and substantial reductions in tradedistorting domestic support. The negotiations on agriculture taking place in the WTO should deliver an outcome that is consistent with the ambition set out in the Doha mandate. Special and differential treatment for developing countries shall be an integral part of all elements of the negotiations and shall take fully into account development needs in a manner consistent with the Doha mandate, including food security and rural development. Non-trade concerns of countries will be taken into account, as provided for in the Agreement on Agriculture, in accordance with paragraph 13 of the Doha Ministerial Declaration.

   ...

   77. Efforts at extending market access liberalization for non-agricultural products under the Doha Work Programme should be intensified with the aim of reducing or, as appropriate, eliminating tariffs, including tariff peaks, high tariffs and tariff escalation, as well as non-tariff barriers, in particular on products of export interest to developing countries. Negotiations should take fully into account the special needs and interests of developing countries and LDCs, including through less than full reciprocity in reduction commitments.
10. Such measures can contribute to strengthening an enabling environment for the progressive realization of the right to adequate food in the context of national food security.

External debt

11. States and relevant international organizations should, as appropriate, pursue external debt relief measures vigorously and expeditiously in order to release resources for combating hunger, alleviating rural and urban poverty and promoting sustainable development. Creditors and debtors must share the responsibility for preventing and resolving unsustainable debt situations. Speedy, effective and full implementation of the enhanced heavily indebted poor countries (HIPC) initiative, which should be fully financed by additional resources, is critical. Furthermore, all official and commercial creditors are urged to participate in this initiative. Heavily indebted poor countries should take or continue to take policy measures required to ensure the full implementation of the HIPC initiative.

Official development assistance

12. Consistent with the Monterrey Consensus, developed countries should assist developing countries in attaining international development goals, including those contained in the Millennium Declaration, by providing adequate technical and financial assistance and by making concrete efforts towards the targets for ODA of 0.7 percent of GNP to developing countries and 0.15 percent to 0.2 percent of GNP to least developed countries. This should be linked to efforts to improve the quality and effectiveness of aid, including through better coordination, closer integration with national development strategies, greater predictability and stability and genuine national ownership. Donors should be encouraged to take steps to ensure that resources provided for debt relief do not detract from ODA resources intended to be available for developing countries. Developing countries are encouraged to build on progress achieved in ensuring that ODA is used effectively to help achieve development goals and targets. In addition, voluntary financial mechanisms supportive of efforts to achieve sustained growth, development and poverty eradication should be explored.

International food aid

13. States that provide international assistance in the form of food aid should regularly examine their relevant policies and, if necessary, review them to support national efforts by recipient States to progressively realize the right to adequate food in the context of national food security. In the broader context of food security policy, States should base their food aid policies on sound needs assessment that involves both recipient and donors and that targets especially needy and vulnerable groups. In this context, States should provide such assistance in a manner that takes into account the importance of food safety, local and regional food production capacity and benefits, and the nutritional needs as well as cultures of recipient populations.

Partnerships with NGOs/CSOs/private sector

14. States, international organizations, civil society, the private sector, all relevant non-governmental organizations and other stakeholders should promote the strengthening of partnerships and coordinated action, including programmes and capacity development efforts, with a view to strengthening the progressive realization of the right to adequate food in the context of national food security.

Promotion and protection of the right to adequate food

15. The organs and specialized agencies related to human rights should continue to enhance the coordination of their activities based on the consistent and objective application of international human right instruments, including the promotion of the progressive realization of the right to adequate food. The promotion and protection of all human rights and fundamental freedoms must be considered a priority objective of the United Nations in accordance with its purposes and principles, in particular the purpose of international cooperation. In the framework of these purposes and principles, the promotion and protection of all human rights, including the progressive realization of the right to adequate food, is a legitimate concern of all Member States, the international community and civil society.

International reporting

16. States may report on a voluntary basis on relevant activities and progress achieved in implementing the Voluntary Guidelines on the progressive realization of the right to adequate food in the context of national food security, to the FAO Committee on World Food Security (CFS) within its reporting procedures.

Index